Technology

and

Society

Rewards and Challenges

Carmel M. Toussaint

Phronetech Writing
ISBN: 978-0-9959098-0-9

Phronetech Writing has no responsibility for the persistence or accuracy of URLs for external or third-party internet websites referred to in this publication and does not guarantee that any content on such websites is, or will remain, accurate or appropriate.

Publisher: Phronetech Writing
Designed by: http://bookdesign.ca/
Edited by: Dania Sheldon, DPhil (Oxford), English Language & Literature

First edition July 2017
Front Cover
The book front cover picture has been chosen in recognition of the performance of Pioneer 10 and 11, the first human-made objects to resist the velocity from the solar system, as part of the Planetary Gran Tour Mission. A gold anodized plaque attached to the antenna of these spacecraft contains information about humankind and the origin of these spacecraft in case they are found in the universe by an intelligent life form. The plaque was designed by the famous astronomers and astrophysicists Carl Sagan and Frank Drake. It is my hope that humankind will always maintain a commanding wisdom on technology, our evolutionary creation.

Back Cover
The back cover illustrates one of the common issues in society: paradigm shift. It is undisputable that information technology, globalization, automation have substantially altered the way of life more visibly in the last twenty five years and undoubtedly in the future. It is also my hope that society will learn how to adapt to this new environment.

Phronetch Writing
14061 - 2408 Lakeshore Blvd West,
Etobicoke, ON M8V 4A2

CONTENTS

Acknowledgments 6

Prologue 7

Chapter 1   A Path to Complexity 13

Chapter 2   An Inevitability 43

Chapter 3   Socioeconomic Impact of Technology . . .83

Chapter 4   Commercialization 119

Chapter 5   Progress: Fiction or Reality 147

Chapter 6   A New Perspective 165

Chapter 7   Epilogue 189

Appendix I

Notes and Glossary 199

Appendix II

Major Inventions of the 20th Century
and the First Decade of the 21st Century 223

Social Progress Index 239

Appendix III

Timeline for the Leisure Society 241

Bibliography & Online Resources 243

Index 251

List of Figures

Figure I.1 Accelerating Growth in Technology 40

Figure II.1 The Spinning Wheel 52

Figure II.2 The Printing Press. 53

Figure II.3 The Diffusion of the Gutenberg Printing Press 54

Figure II.4 The Steam Engine 56

Figure II.5 Penetration of Bt Cotton in Farming in India 67

Figure II.6 Calorie Consumption by Adopters of Bt Cotton . . . 67

Figure II.7 Life Expectancy of the World Population. 69

Figure II.8 World Energy Consumption by Fuel 73

Figure II.9 World Energy Consumption. 74

Figure III.1 Urban Population 92

Figure III.2a A Mansion in the Early 1880s 93

Figure III.2b Child Labor 94

Figure III.3 A Smoke Pollution in the English Town of
Widnes, Late 19th Century 95

Figure III.4 Molecular Imaging & Therapy 100

Figure III.5 Potential Benefits of Nanotechnology
—Health Sector 102

Figure III.6 Growth of Nanotechnology. 103

Figure III.7 Distribution of Internet Users in the World 107

Figure III.8 Percentage Change in the World's Population by Age:
2010–2050. 112

Figure III.9a Total Populationby Variant 113

Figure III.9b Total Population by Major Age Groups 113

Figure IV.1 The Crystal Palace 120

Figure IV.2 Industrialization 123

Figure IV.3a Labor Productivity in the Business Sector 132

Figure IV.3b Productivity in Advanced Economies 133

Figure IV.3c U.S. Productivity in the Non-Farm Business Sector 134

Figure IV.4. Declining Share of Young Firms 136

Figure IV.5 GDP in Singapore 2007–2014. 142

Figure VI.1 Communication Modes 170

Figure VI.2 Concentration of Carbon Dioxide over Time . . . 185

Figure VII.1 Human Evolution 189

List of Tables

Table III.1 Promises of Nanotechnology. 101

Table IV.1 Production of Potatoes in Prince Edward Island . . . 131

Table IV.2 The Global Creativity Index—Overall Rankings . . . 139

Table V.1 World GDP by Year 150

Table V. 2 Canada- USA GDP 2008 – 2013 150

Table V.3a The Social Progress Index 153

Table V.3b The Social Progress Index Result 2015 154

Table VI.1a World Internet Users in 2015 171

Table VI.1b World Internet Users in America in 2015 172

Acknowledgements

The issues dealt with in this book are not new. Many essays, articles and books have addressed them, and surely more are to come as time goes by. However, it is my hope that readers find the views expressed in these pages to be different and beneficial, and that these views will generate discussion on the multiple aspects of technological development facing us all.

I would like to emphasize the tremendous amount of work that has been done thus far by various universities, transhumanist associations and social media participants to democratize technology and its potential. More specifically, it would have been difficult, if not impossible, to write this book without the books and articles of several philosophers, scientists and futurists, such as Ray Kurzweil, Nick Bostrom, João Pedro de Magalhães, and many others to whom I deeply apologize if their names are not mentioned. The writings of these scientists have been inspirational. That said, any error is mine, and I take full responsibility for the content of this book.

I also would like to express my profound gratitude to my editors, Dania Sheldon and Nancy Wills, for their arduous exercise of transforming the manuscript into an easy-to-read document, and my reviewers for their abundant and constructive comments. Special thanks are also due to the various organizations and authors that have given permission to reproduce parts of their work. Finally, but not least, I dedicate this book to the general public, unaware of its position in the driver's seat and still in search of a silver lining in technology.

Homo sapiens display the resilience of a stubborn boxer whose efforts are only encouraged by each knockdown.
—Source unknown

Prologue

"It is the obvious which is so difficult to see most of the time. People say 'It's as plain as the nose on your face.' But how much of the nose on your face can you see, unless someone holds a mirror up to you?"
—Isaac Asimov, *I, Robot*

This book has emerged from my reflections about the current state of technology, the associated benefits and challenges, and their related controversies. Intrigued by a lack of perspective on this matter, I started digging more and more, and I meticulously reported the results of my findings in a document that I called "My Journal." During this process, I made a few interesting and motivating discoveries. First, we human beings are very special social creatures, and the power of our collective learning has propelled us to where we are today. Second, technology, our most precious creation, is an expression of our volition to live; the same phenomenon, to some extent, can also be seen throughout the animal kingdom. Third, we are living in a very important period of our history on this planet; one in which we are making transitions to new paradigms caused by the inventions and innovations affecting our work, our social habits, and even our political system. Eventually, in the process of consolidating the impact of technology on society now and in the coming years, this journal took on a life of its own, leading to this book.

The way of living is in constant mutation, and the future for some is fervently appealing, while for others it is either unimportant, negatively suspicious, or fearsome. Every century has brought to humankind its load of pleasant technological surprises. The invention of computers and, more recently, the development of the internet and social media, were the goodies of the last and current centuries. Wherever a person is on planet Earth, enormous benefits can be drawn from at least two of the aforementioned innovations, from the wealthiest to the poorest countries of the world.

The 21st century brings even more promises for the betterment of life on our planet. In the face of the magnitude of the task at hand, which is to shed light on current and future impacts of technological innovations, I am reminded of the revolution of information diffusion during the Renaissance, when Gutenberg introduced the movable type printing press in 1450. As a new invention, it had to be brought to the writer's awareness and to the public. That invention has substantially changed the printing industry by creating some standardization, thereby facilitating greater access to data and information. Although no information is available on that marketing effort, it is fair to suspect that it was quite a challenge. Today, the situation is not much different in terms of inventions' scope and potential ramifications. Current technological inventions and future innovations have become emotionally charged and induce different feelings across the emotional spectrum, from deep indifference to the utmost enthusiasm.

Nonetheless, literature and movies about the future are quite abundant. Thanks to Hollywood, our consciousness is well seeded with images of various space crafts, life forms encountered during interplanetary trips, invasion of our planet by "aliens" to secure their own future, or anticipation of events that are remotely possible. This is not to say that I am against speculation. It often is conducive to scientific experimental work that may support or deny the said speculation. Perceptions of the future, although apparently very uniform and widespread, have in fact permeated three different strata of society in three different ways.

In the first stratum, which encompasses the largest segment of the population, indifference seems to be the proper qualifier when it comes to inventions or future innovations, because of their technical nature. Discussions or conversations with friends about the future would in all likelihood involve unpopular technical terms and concepts that are at the root of this indifference. Furthermore, the potential changes appear to be so farfetched that people see them as either pure fantasy or wishful thinking and therefore not credible.

Reluctance characterizes the second stratum. This segment of the population tends to be more educated but are reluctant because so many future predictions have not materialized, such as Y2K in terms of computers' incapacity to differentiate the 1900s from the 2000s leading to catastrophic problems and other similar irrational predictions. In addition, and this seems to be common to the above groups, religious beliefs seem to be a latent force that filters potential societal benefits in a very radical manner. Examples of this mindset can be found in people's opinion about the contraceptive pill, blood transfusions, implants, cryonics, and so on. These issues will be addressed later, in Chapter V, as part of the ethics section.

The third stratum, also an educated group, tends to be supportive, curious and, to a certain extent, even enthusiastic about the forthcoming changes resulting from technological development within their value system, whatever this might be. However, this is not to say that the support or rejection of technology is entirely based on cognitive factors. It is also well accepted that affective factors such as fear and perceived risks of negative outcomes play an important role in people's attitudes towards technology and, moreover, these affective factors can overrule any quantitative analysis.

Discussions and lectures about technology and future life are gaining momentum around the globe, particularly among the intellectual elite and the third stratum. Some allegations or positions are more direct and descriptive than others, each with different perspectives. However, taken together they demonstrate the widespread attention that the future is receiving. In Canada, a poll conducted in 2014 by Ipsos Global Trends suggested that 51% of Canadians feel they need modern technology because only this can help to solve future problems. South of the border, according to a survey conducted by the Pew Research Center and the *Smithsonian* magazine

(April 17, 2014) about America's views of technology and the future, 59% of Americans reported being optimistic about the coming technological changes. However, the results also suggest that the population remains divided on other issues, such as driverless cars, brain implants, and longevity.

At this stage of technological development in which totally new products and services are daily coming to the market, it is fair to say that our modus vivendi (lifestyle) will not stay the same. Like the unstoppable spread of a tree's roots, these new products and services have been and will continue to affect all aspects of our lives, from both a societal and an economic standpoint. While beliefs affect choices, which in turn affect action, one may choose to ignore this paradigm shift or be predisposed to it. Those who adopt the former position will have robbed themselves of the opportunities that the evolution of technology will bring forward as never before. Those choosing the latter position will be more informed of the risks and benefits ahead but also will participate in this new development. It is worth mentioning that we will be at the steering wheel, driving our future by the very choices we make and our adaptation thereafter. I do recognize that this statement is quite a simplified explanation of the interplay of cognition, emotion and the resulting social behavior towards technology, but this is the reality of the challenges ahead of us all.

My purpose in writing this book is threefold. First, in Chapters I and II, I aim to clarify the very nature of technology by explaining how we got where we are today. I perceive that this development is met with indifference, reluctance, or rejection from the majority of people due to the information with which they are presented. In this book, I strive to use easy-to-understand, layman terms as much as possible. It is my hope that the notes and glossary offered in Appendix I will help with this objective. Second, I aim to outline the practicality, convenience, benefits, and potential dangers of new products and services and the meaning of those changes for our society. It is my position that from the beginning of our sedentary society, or even far earlier, up until today, technology has been seen successively as a servant, friend, and ally. As in the biological sphere, there is an obvious development from infancy to adulthood. Exploring these ideas is the purpose of Chapters III to V. Third, it is important to understand that we all will have to make important and delicate choices. Some may perceive certain applications of the emerging technology to be invasive and see them as affecting the fabric of who we are. Perhaps the virtues and values that we embrace or that are imposed by our culture may need to be revisited in light of the current environment so that we can look at the future from a different perspective. As most of our dearly held values will be challenged in the next decades, we will need to rethink what keeps us together as a society. The recurring question, despite the benefits of technology, is whether we have really progressed as a society. It is relatively easy to conflate the use of scientific advances with moral progress. Although the idea of progress means different things to different people, based on cultural, political, or economic differences, I believe that there are some universal values common to all of us on which we can find some agreement. Examining these is another purpose of Chapters III to V. Chapters VI and VII address the question of where we are heading with all of these issues, including our current socioeconomic challenges. For the sake of conciseness, I have included at the end of each chapter some suggested readings to complement the information provided. The number in parenthesis indicates the reference(s) in the suggested reading section.

The use of behavioral attributes to characterize the relationship between technology and its human creators is a challenging and audacious task, not least because of the complex nature of technology, the finite horizon of humankind, and the changing positions about what is wanted. Many support the idea that technology, our unique creation, grows and sustains our own development. Along this line, I have argued that technology has followed an ascending and dynamic path in its relationship with us: from the first stone tools to the Renaissance to being a servant; from there to the Second

Industrial Revolution to being a friend; from the technological and digital revolutions of the 20th century to being an ally up to the present today. It is my position that this inseparable and invisible relationship will continue to evolve to a degree that we cannot fathom because of the unpredictable nature of both parties. In one of his bestselling books, *What Technology Wants*, Kevin Kelly, the maverick of *Wired* magazine, refers to technology as the Seventh Kingdom to underline the uniqueness of technology and its innateness in human beings. I support this creative approach. Indeed, the evolution of technology predates the emergence of language in our species. But, I also see technology as the expression of our will to live. From this perspective, the will is in constant evolution to improve our lives.

Reading this book, scientists in any specific discipline may feel frustrated with the limited amount of information given to substantiate specific issues in their particular field. The fact of the matter is that this book is dedicated to the general public, although by the nature of the subjects addressed, I must admit that the reader should have some background information on sciences in general. As mentioned earlier, I have tried to fulfill this requirement by providing additional sources for interested readers at the end of each chapter and a glossary in Appendix I.

In the spirit of Ursula K. Le Guin, who wrote about communication in her superb book *The Wave in the Mind*: "Words do have power. Words are events, they do things, change things." The ultimate goal of this book is to raise awareness of the opportunities and challenges facing us and to facilitate discussions about these issues. At a bare minimum, if at the end of this book the reader starts asking questions, this objective will be met.

CHAPTER I

A Path to Complexity

"Science is organized knowledge, wisdom is organized life."

—Immanuel Kant

If there is something unique about humankind, it is our quest for constant improvement of life. It is my position that this thrust is an innate quality of our species responding to external pressures, as witnessed by the history of Homo sapiens and its technology. The debate about technology is now quite intense, and I expect it to become more inflamed as more innovations penetrate the market. All the trend reports point to the second decade of the new millennium as a period of increased efforts to merge technology with human intelligence. Nothing can be closer to the truth, as evidenced by the recent developments in informatics, artificial intelligence, and nanotechnology.

Before going any further, let's ask ourselves why we are at this stage of great technological advances today in our civilization. To be able to answer this question, we must reflect on the essence of both science and technology. A good starting point would be to look at the definitions of both. According to the International Technology Education Association (ITEA), "science deals with and seeks the understanding of the natural world and is the underpinning of technology," while technology is the process by which humans modify nature to meet their needs and wants. In other words, technology in a broad sense is the expression or manifestation of the will of our species to live at an increasingly improved level. The relationship between the two is quite evident. One cannot exist without the other. Science supports technology by pushing the boundaries of existing knowledge, and technology enriches the research field of science through bringing about challenging innovations to meet society's needs and wants.

But if, according to the above definition of technology, humans modify nature to meet their needs and wants, the race to perfect our world becomes clear. The thrust is both innate and societal. Needs are quite different than wants. I might need a car to go to work, but I may also want the devices of a luxury model, which could be beyond my budget. I recognize that the notion of luxury features is relative. For instance, not too long ago, air conditioning, power steering, and automatic windows were upgrades to be bought as options. Today, they are part of most basic car models. These improvements, made affordable by technological developments and competitive market forces, occurred as a result of our increasing needs and wants. This is to say that while our needs are factually limited or imposed by our budget, our wants are limitless. I must add that no value judgement is formulated at this point as to whether our insatiable epicurean appetite for novelties is desirable or not.

To study the nature, consistency, and speed of satisfying our wants and needs, let's review some of the technological innovations from the end of the Middle Ages to the first decade of the new millennium. But first, certain clarifications need to be made, which will be useful for understanding the position I take in this book:

1. The review period starts from the Renaissance, which for our purpose covers the period from AD 1400 to 1600. The separation of the Middle Ages from the Renaissance period greatly varies between authors, depending on the purpose and the scope of a study. For example the Renaissance is deemed to have occurred from 1400 to 1688 in Italy; from 1494 to 1610 in France, with a 28-year peak between 1515 and 1547; and

from 1485 to early 1600 in England. My rationale for selecting the timeframe mentioned above stems from the fact that it facilitates an understanding of the subsequent periods.

2. The reported date of an invention is sometimes confused with the date at which that invention was patented (since patenting came into existence). This sometimes creates a distortion in the reporting, particularly if the invention never gets patented. Nevertheless, it is recognized that the patent date plays an important role, among other things, in the study of the history of technological development.

3. Invention is different from discovery (2). Invention results from creative thinking, that is, a totally new product or process that achieves a breakthrough, be it radical or not; the invention of the transistor, for example, was a radical breakthrough. The invention of the hula hoop, in 1958, was not, despite its popularity at the time. Discovery is the identification or recognition of an existing phenomenon, process, or element because of advances in scientific knowledge. An example of discovery is the identification of oxygen, the most abundant element in the Earth's crust; it was discovered by the Swedish chemist Carl Wilhelm (1742–1786). Credit for this discovery must also be given to the English chemist Joseph Priestly (1733–1804). However, they worked independently. Both discoveries occurred at the same time, in 1774; this example illustrates the concept of simultaneous invention and discovery.

4. I am taking this opportunity to differentiate invention from innovation. Innovation is the application of an invention, the introduction of a new product or process to the market. Innovation is market, profit driven, the manifestation of the current or anticipated consumer's wants and needs, driven by commercial gain or other goals. On this latter note, no one, for instance, would now launch a buggy-whip business. However, a programming school or a computer maintenance

service business would be far more financially promising. From a societal perspective, invention, discovery, and innovation provide a panorama of humanity's knowledge and determination to achieve some value in the context of life improvement.

The European Renaissance is an interesting period, characterized by the re-awakening of intellectual thought subsequent to the Middle Ages. In that earlier period, religious and political wars suppressed creative and intellectual activity. However, this is not to say that no invention and innovation occurred during the Middle Ages, particularly in the construction, mining, and maritime transportation sectors, facilitated by Roman and Byzantine antecedents and trading networks with the Islamic world, China, and India. As mentioned earlier, the Renaissance occurred at different times in different European countries. For the purpose of this book, England and France will be the focus. Thus, I have designated the Renaissance period as being from AD 1400 to 1600.

From the period of the ancient Chinese civilization (2100 BC) to the fall of the Eastern Empire (AD 1453), there were many inventions, innovations, and discoveries that changed human lives and continue to affect our lives to the present day. Despite the loss of one-third of Europe's population as a result of the plague known as the Black Death in 1446, in the late Middle Ages, our species has shown the resilience to renew itself and re-emerge as never before. Technology could be seen as a survival instrument, a faithful servant in this period, not from a master–slave perspective but from a symbiotic one: a relationship in which both parties benefits from helping each other. From the axe maker of the Paleolithic period to Gutenberg's movable-type printing press circa AD 1450, the creation process was compartmental and oriented towards the fulfillment of mostly primary needs. This is reflected in the birth of implements for agricultural, medical, and other day-to-day living purposes.

A compilation of humans' inventions, innovations, and discoveries can be found in the outstanding book of the late Isaac Asimov, *The*

Chronology of Science and Discovery, which goes as far back as 4000 BC. I was pleasantly surprised to note that we have inherited many inventions from previous civilizations, particularly China. This may sound trivial, but even our daily coffee beverage was discovered in Ethiopia in AD 850 and became a popular drink in Europe in the 12th century. Although our review period starts at the Renaissance, certain pre-Renaissance inventions need to be mentioned because of their impact. A few examples are:

- Gun powder*   ~ 1249
- Eye glasses   ~ 1290
- Spinning wheel*   ~ 1298
- Mechanical clock*   ~ 1300

Renaissance

Similarly, many inventions of the Renaissance can be described as being not only practical but also enduring, as their impact continues to be felt today, as will be seen for others in subsequent periods:

- Printing press   ~ 1436
- Artificial limbs   ~ 1543
- Microscope   ~ 1590
- Principles of heart and
  blood circulation   ~ 1600

The above is not a comprehensive list of inventions during the (pre)Renaissance period. There are many other sources that list the vast array of inventions not illustrated in this book. I suggest reading Asimov's *The Chronology of Science and Discovery* because of the analytical robustness of the subjects covered.

---

* From a chronological perspective, the term "Kingdom of England," encompassing Wales, Scotland, and Ireland, can be used for the period up to May 1, 1707. The appellation Great Britain came into practice on January 1, 1801, when Ireland entered the Kingdom as per the Act of Union. The name United Kingdoms of Great Britain and Northern Ireland was established by the Royal and Parliamentary Titles Act of 1927. Presently,the abbreviation UK is meant to identify a country that includes England, Scotland, Wales, and Northern Ireland.

Three of the inventions in the above list, marked with an asterisk (*), are worth underlining because of their social impact, which will be fully discussed in Chapter III: gun powder, which changed the way war was fought; the spinning wheel machine (from China), which moved English textile production from to artisanal stage to the pre-industrial stage; and the mechanical clock, which measured time in a systematic way by allowing time to be compressed, sliced, and expanded for work efficiency and effectiveness.

But before we go any farther, another observation needs to be underlined. A good 50% of the inventions listed above were created in China, starting with gun powder in AD 1044. Moreover, many other inventions not listed above predate the Renaissance and reflect China's technological strength, such as the inventions of the compass in 221–206 BC, paper in AD 105 during the Eastern Han dynasty, metal stirrups in AD 300, the wheelbarrow in AD 400, as well as silk fabric and porcelain in AD 700.

This then raises the question as to why China's technological superiority did not overtake Western European technology, particularly that of England.* This historical fact, known as the Great Divergence (a term coined by the political scientist Samuel P. Huntington in 1996), has drawn the attention of many scholars. Asimov offered two reasons to substantiate his position on this matter.

The first was the lack of reliable information because of the decision of the first emperor of the Ch'in dynasty, Shih Huang Ti (259–210 BC), to burn all books (except those in practical arts), to free the land from its past, thus eradicating evidence; the second was that scientific advancements can only be judged by their effects on the world: "*Discoveries, innovations and inventions can only count when they affect society.*"

According to Asimov, in the first decade of the 15th century, during the reign of Emperor Yung-lo, the Chinese launched a series of expeditions in the Indian Ocean, conquered the Indonesian islands, and defeated the ruler of Ceylon. However, again according to Asimov, China was a world unto itself; successive emperors had no interest in dealing with other nations, eliminating its chance at world influence and leaving this to other, much smaller, much weaker, and much less advanced nations.

A total reversal occurred, particularly in the mid-20th century, and China became a world power as we know it today. This situation reminds me of one of the theories of Montesquieu written about by Alan McFarlane in his book *The Riddle of the Modern World: Of Liberty, Wealth and Equality*: "If all necessities could be produced within a boundary, this would lead to despotism. In other words, commerce encourages freedom. If commerce is the cause of freedom, it is also the consequence of commerce. It can only flourish where there is a certain freedom." This quote is bolstered by a passage in *Guns, Germs, and Steel*, by Jared Diamond: "Once China was unified in 221 BC, no other independent state ever had a chance of arising or persisting for long in China... Thus, geographical connectedness and only modest internal barriers gave China an initial advantage... but China's connectedness eventually became a disadvantage, because a decision by one despot could and repeatedly did halt innovation. In contrast, Europe's geographic balkanization resulted in dozens or hundreds of independent, competing statelets and centers of innovation." Jared Diamond attributes European technological development to the result of the Age of Exploration of the 15th and 16th centuries, the Scientific Revolution of the 16th and 17th centuries, and the Industrial Revolution of the 18th and 19th centuries.

But there are also additional factors, or rather combinations of factors, explaining the Great Divergence between the East and West. Among the abundant papers on this subject, world historians propose various explanations for the cause of the Industrial Revolution in the second half of the 18th century in England: (1) richness in raw materials at the center of the Empire and its peripheries; (2) the amount of wealth earned from the colonies and worldwide trade, and the investments of wealthy men in machinery and technology; (3) the geographical location that put England away from the harmful wars of continental Europe; and (4) the existence of a stable government that supported the efficiency of institutions and encouraged scientific development. In my view, this last factor should be given special consideration in the analysis of the different paths taken by the two countries.

As can be seen, no single factor can justify the Great Divergence; rather, it was due to a combination of all them. Shamkhal Abilov, Senior Research Fellow at Qafqaz University, in Azerbaijan, came to a similar conclusion in his paper entitled "The Great Divergence between China and England: Why Industrial Revolution Happened in Europe?" Neither the culture and religion of Europe nor the economic history of the world sufficiently explains the Great Divergence.

Every nation in the course of its history has faced shifting circumstances, defining moments when politico-economic decisions or the lack of them generate losses for some and gains for others. Today, globalization and information technology together offer opportunities and challenges for maintaining technological superiority in that their interplay facilitates the acquisition of knowledge and skills at a faster pace. In addition, the mobility of capital allows a firm to take advantage of market opportunities wherever they are available, to maximize profit. It need not take long for a host country to become a production leader in a firm's home country, particularly if the appropriate skill level and business incentives are in place. In today's fast-paced, nerve-racking, and competitive environment, a firm can maintain its technological superiority only by coming up with new products and services or manufacturing processes, resulting from its own research and development initiatives. Conceivably, this is also valid at the macro level. A country can maintain its technological superiority if high priority is given, inter alia, to basic and applied research, the source of inventions and innovations. Market and government cannot be disassociated. While the former is in perpetual change in a free-market economy, the latter has a role to play in keeping a level playing field for all.

The 17th Century

"We owe a huge debt to Galileo for emancipating us all from the stupid belief in an Earth-centered or man-centered system. He quite literally taught us our place and allowed us to go on to knowledge."
Christopher Hitchens

"All our knowledge begins with the senses, proceeds then to the understanding, and ends with reason. There is nothing higher than reason."
Immanuel Kant

"It is not what the man of science believes that distinguishes him, but how and why he believes it. His beliefs are tentative, not dogmatic; they are based on evidence, not on authority or intuition."
Bertrand Russell

The 17th century shows some momentum in technological developments. Innovations start becoming part of a creating culture. Indeed, the number of innovations grew substantially during the 17th century. Most remarkably, the recognition of the terms science and "scientists," who in the preceding period were called philosophers, was just beginning. René Descartes, Blaise Pascal, and Isaac Newton, for example, were recognized as notable scientists. The statement of Mary Bellis in her blog best describes the academic environment: "By the end of the 17th century a scientific revolution had occurred and science had become an established mathematical and empirical body of knowledge." There is no unanimity on the periodization of the Scientific Revolution. Most historians seem to agree on the period starting with the formulation of the heliocentric theory of Copernicus (1550) and ending with the publication of Universal Laws and the The Mechanical Universe by Isaac Newton in 1687.

Among the most important inventions and innovations of this period were:

- Earliest human-powered submarine   ~ 1608 by Cornelius Drebbel
- Telescope*   ~ 1608 by Hans Lipperhey
- First blood transfusion*   ~ 1625 by Jean-Baptiste Denis
- Steam engine   ~ 1629 by Giovana Branca
- First intravenous anaesthetic injection   ~1656 by Christopher Wren
- Calculating machine   ~ 1679 by G.W. Leibniz
- Steam pump   ~ 1698 by Thomas Savery

From the above, two inventions are worth elaborating upon: the telescope and the first blood transfusion. The enormous influence of the steam engine will be discussed later because of its contribution to the Industrial Revolution. The telescope stands as one of the greatest inventions of the 17th century in the context of understanding our universe. It was invented by Hans Lipperhey in 1608. The history of the invention of what is known today as the telescope is unclear. Some pretend that the original idea came from Lipperhey's workers. Others believe that it came from his work experience. Nevertheless, the German-Dutch spectacles maker is generally recognized as the inventor of the instrument of "seeing things far away as if they were nearby," the original name of the telescope. However, Lipperhey did not get a patent from the States General of the Netherlands because of a similar application at the same time. He was, though, largely rewarded by the Dutch government for copies of his invention.

It was through the telescope that Galileo Galilei settled the controversy between the proponents of the geocentric system (since Aristotle and Ptolemy) and the followers of the heliocentric system, through instrumental observation. Galileo, one of the most important scientists of the 17th century, put forward many theories, including the theory of uniform acceleration; all bodies, regardless of their weight, fall at an equal rate in the absence of friction. Although he cannot be credited for the creation of the telescope, he did increase its performance magnitude by 32 times.

Preceding Galileo, the publication of the discovery of Nicolai Copernicus (*On the Revolution of Heavenly Bodies*) stated that the Earth rotates around the sun and not the other way around, which had a major effect on the belief system of the time. The Roman Inquisition fundamentally opposed the rationale of the heliocentric theory and even prohibited this information from being published because it was contrary to the Bible's teachings, as can be found in very literal interpretations of Psalms 93:1, 96:10, and 104:5, and Ecclesiastes 1:5. Galileo substantiated his position in one of the most articulate documents about the separation of science from theology: "I believe that natural processes which we perceive by careful observation or deduce by cogent demonstration cannot be refuted by passages from the Bible." The full letter of Galileo to the Grand Duchess Christina of Tuscany (7) is referenced at the end of this chapter.

The discoverer of heliocentric gravitation had a terrible fate. Galileo was accused of heresy and confined to house arrest without visitors for the rest of his life. He died there of old-age-related diseases. His theory, quite rightly, still prevails, and he is often referred to as the father of astronomy. The democratization of information by the Gutenberg printing press disseminated Galileo's theory throughout Europe. But it was only in 1835 that officially the Church lifted its ban.

Technology, innovations, and discoveries cannot be suppressed. I have chosen to briefly mention this case at this particular moment to underline the constraining role of the Church and the State in scientific issues. The notion of free speech is still an issue today, particularly when new ideas shake the foundations of political, social, and religious systems.

Jean-Baptiste Denis, French mathematician and philosopher, performed the first blood transfusion on June 15, 1667, from a sheep to a 15-year-old boy. This was known as a *xenotransfusion*, that is, when the blood is transfused from an animal to a person. The procedure required introducing the carotid artery blood of the animal to the human through a vein in the patient's inner elbow. A total of four patients were treated with blood transfusions by Denis. Of these, two survived and two died. Based on the information available, it would appear that the preexisting conditions of the last two were not as "good" as the first two. With the wisdom of time, we can all appreciate the revolutionary character of this particular intervention. The third patient survived three transfusions, but upon his death the family accused Denis of malpractice and he was ordered to stop this type of intervention unless specifically permitted by the Faculty of Medicine of Paris. Then the practice of blood transfusion faded out as suddenly as it began. This was quite a revolutionary procedure at that time, and needless to say, it became very controversial. In fact, although partially successful, xenotransfusion was banned in 1670 by the French Parliament, the English Parliament, and the Pope until 1901, when the biologist and physician Karl Landsteiner discovered the four main blood groups.

I must also underline the substantial medical contributions of the preceding century in surgery by Ambroise Paré, and in anatomy and blood circulation by Andrea Vesalius and William Harvey. Their work constituted what can be called the Renaissance in the medical field. Their work also signaled the diminution of the Church's power in science in that many studies could be made on dead human bodies instead of animals, as had previously been prohibited by the religious authorities.

It would be unjust to move to the 18th century without outlining the contributions of Francis Bacon and René Descartes not only to the method of scientific enquiry but also to the development of knowledge. With the introduction of the telescope and the microscope, humanity no longer relied on the unaided physical senses but on instruments for objective observations. Francis Bacon said that a true science progressed "in a just scale of ascent, and by successive steps not interrupted or broken, we rise from particulars to lesser axioms; and then to middle axioms, or above the other; and last of all to the most general." This inductive method had to be applied to all concepts and ideas.

But most importantly, there is a quote from Bacon on his view about technological advancements which is worth mentioning. He saw humanity's tireless efforts to conquer nature as a divine mandate. Referring to Genesis 1:28 in *Novum Organum*, he wrote that "man by the fall, fell at the same time from his state of innocence and from his dominion over creation. Both of these losses, however, can even in this life be in some parts repaired; the former by religion and faith, the latter by the arts and sciences." Bacon was one of the first writers of the Renaissance to discuss the impact of technology on society with the view to improving life's conditions through inductive reasoning. He clearly indicated the rationale of the thrust of science and technology. I can then suggest that he is one of the precursors of transhumanism because of his conviction of the overwhelming duty conferred upon humankind to improve life and conquer natural limitations.

Like Francis Bacon, René Descartes spent a considerable amount of time researching and defining a scientific method for approaching and solving problems, key to scientific discovery. Contrary to Bacon, his method was deductive in that it began with the general, from which successive non-substantiated components were eliminated, to arrive at the truth. Descartes is also considered to be the father of modern philosophy because of the dualistic dimension of his philosophical tenet "Cogito ergo sum" (I think, therefore I am), one of the key tools for testing self-awareness involving both the mind and the body.

The development of science owes much to Isaac Newton for introducing the universal laws of motion. For him, there was one kind of matter, one set of laws, one kind of space, and one kind of time. We owe to Newton a new paradigm about the nature of things and their cause-and-effect reality. His laws of motion were revisited by Albert Einstein in the 20th century via the concept of curved space and time, which more fully explains the concept of gravity. Perhaps this is the reason why science works so well and is always on an ascending trend. There are no sacred truths.

It should be noted that towards the end of the 17th century, a framework started taking place for furthering technological development. In 1662, the Royal Society of London (5) was founded, followed by the founding of other scientific societies. These official entities assumed the position of reviewing study papers and discussing new theories. This openness was embraced centuries later by the astronomer Carl Sagan: "At the heart of science is an essential balance between two seemingly contradictory attitudes and openness to new ideas, no matter how bizarre or counterintuitive they may be, and the most ruthless skeptical scrutiny of all ideas, old and new. This is how deep truths are winnowed from deep nonsense."

The 18th Century

The 18th century is commonly called the Century of Reason or the Enlightenment, in which period the First Industrial Revolution occurred. I believe that if a time traveler were stopped in this period, he/she would notice a different way of thinking, a new era in human development. The number of innovations had substantially increased, revealing an astonishing trend showing the beginning of the growth of science, the foundation of technology. The environment was propitious for the Industrial Revolution because of the number of inventions, and the growing knowledge and skills of the previous centuries. This accumulated intelligence was passed on to the new generation, supplementing the thinking of preceding century. For the first time, industrially built machines such as the steam engine and the spinning wheel facilitated the mass production of goods. This was the beginning of a system of producing more with less human labor and with less excruciating physical fatigue being experienced by both laborers and animals in the production process.

The number of inventions continued to increase in scope and impact. The most prominent inventions of this period were the following:

- Encyclopedia   ~ 1750–1782
- Spinning Jenny   ~ 1764 by James Hargreaves

Steam pump	~ 1764 by James Watt
Power loom	~ 1785 by Edmund Cartwright
Electric telegraph	~ 1774 by Georges Louis Lesage
Bifocal eye glasses	~ 1780 by Benjamin Franklin
Steam wagon	~ 1769 by Nicolas-Joseph Cugnot
Smallpox vaccine	~ 1796 by Edward Jenner
Domestic gas lighting	~ 1799 by William Murdoch

First in the above list is the *Encyclopedia*, which was also called the *Systematic Dictionary of the Sciences, Arts and Crafts*, published in 35 volumes between 1751 and 1780, and republished in 166 volumes between 1782 and 1832. This masterpiece of the Enlightenment period recorded accumulated knowledge for future generations. Rousseau, Voltaire, Montesquieu, Diderot, and many others contributed to the redaction of the *Encyclopedia*. Nobody other than Denis Diderot, the main contributor, can better express the purpose of this invention:

> *The goal of an "Encyclopédie" is to assemble all the knowledge scattered on the surface of the earth, to demonstrate the general system to the people with whom we live, and to transmit it to the people who will come after us, so that the works of centuries past is not useless to the centuries which follow, that our descendants, by becoming more learned, may become more virtuous & happier, & that we do not die without having merited being part of the human race.*

It is worthwhile emphasizing that the publication of the *Encyclopedia* was banned for a period of time, like the Gutenberg printing press, like Galileo Galilei's publications. These tangible records of knowledge were perceived as a challenging force by the then religious authorities. The history of the dissemination of information is filled with impediments and restrictions created by a few at the expense of many. The *Encyclopedia*, regardless of its noble objective, was not exempt.

The 19th Century

The 19th century can be considered the ignition that sparked the explosion of human ingenuity.

In the 19th century, there was a phenomenon that has continued to the present day: the metamorphosis of the human–machine relationship. The creation of machines making parts for other machines led to the notion of parts being changeable. This also recognized the physical limitations of humans to meet precise machine specifications. This was a major step in technological development because at this point, the expression of the human will to live materialized into a machine–friend relationship. One can see a shift in the creation process: a certain degree of independence yet amenable to a closer relationship because of the enlarged scope and substance of capacity. This bonding, for lack of a better term, will continue to the next millennium. Among the most important inventions of the 19th century were the following:

- Battery   ~ 1800 by A. Volta
- First electric light   ~ 1809 by H. Davy
- Electromagnet   ~ 1825 by W. Sturgeon
- Electrical circuit   ~ 1827 by G. Ohm
- First computer   ~ 1837 by C. Babbage
- Rubber vulcanization   ~ 1839 by C. Goodyear
- Fuel cell   ~ 1839 by Sir William R. Grove
- Telegraph   ~ 1843 by S.B. Morse
- Stent   ~ 1847 by Charles Thomas Stent
- Sewing machine   ~ 1851 by I. Singer
- Bessemer converter   ~ 1855 by Henry Bessemer
- Tungsten steel   ~ 1868 by Robert F. Mushet
- Telephone   ~ 1876 by Alexander G. Bell
- Electricity   ~ 1882 by G. Ohm
- Coca-Cola   ~ 1886 by J. Pemberton
- First gasoline tractor   ~ 1892 by John Froelich
- Invar   ~ 1896 by C.E. Guillaume

Hundreds of new products were invented and commercialized in the 19th century. Because of discrepancies between many sources, it is difficult to make a specific quantitative statement on the number of inventions and innovations. But as a result of this technological innovative surge, an economic outburst occurred at the end of that period, echoed by many writers. The Czech-Canadian scientist Vaclav Smil, Professor Emeritus in the faculty of Environment at the University of Manitoba, Canada, called 1867–1914 a period of synergy because of the significance of the innovations and inventions that occurred.

The telegraph became popular mainly after the Great Exhibition of 1851 (an extraordinary display of human achievement discussed in Chapter IV). Within three years, Great Britain was wired with a network that transmitted many thousands of messages per week. In the United States, Samuel F.B. Morse invented his telegraph system and connection capabilities with the financial assistance of Congress in 1843. On May 24, 1844, Morse was able to send the following message from Washington to his friend Vail in Baltimore: "What hath God wrought." By 1851, through Morse's perseverance, 50 telegraph companies were operating in the United States. Subsequently, effective written interstate communication was a fait accompli, strengthening communication bonds between people and linking business communities to far-away customers.

Among the most important technological inventions and innovations of this period were in the transportation industry (Goodyear's rubber vulcanization processes and railroad), the clothing industry (the Singer machine and textiles), the communication industry (the telegraph and then the telephone), and the energy industry (battery and tungsten steel).

One can appreciate today the importance of the inventions of the first (mechanical) computer, the electric light, the gas-powered fuel cell, and Invar. The first mechanical computer engine was designed by Charles Babbage in 1837, but it was built for demonstration in 2002—more than a century and a half later (6). Further innovation in fuel cells would bring a solution to the polluting gas emissions of automobiles. Invar is a nickel–iron alloy that undergoes minimal expansion or contraction due to temperature changes. The resulting Invar foil, because of its properties, is used in the production of printed circuit boards.

Some of the above inventions had many contenders because of omissions in the patent applications or financial hardship. For instance, Alexander Graham Bell is credited as the inventor of the first practical telephone in 1876. Five years earlier, Antonio Meucci had filed a patent caveat, but he omitted to include in the description of his invention the key feature: "electromagnetic transmission of vocal sound." Furthermore, his patent, which was valid for one year, was not renewed because of financial hardship.

With the inventions of many agricultural tools throughout the 19th century, culminating in the first gasoline tractor in 1892, U.S. agriculture became increasingly mechanized and commercialized. This mechanization has made the United States the foremost food producer in the world. In addition, the development of machines in the textile and steel industries led to the rise of mechanization and the factory system. The 19th century was also a period of rapid expansion in London. In the transportation sector, the London and Westminster bridges were both operational. In addition, by the last decade of that century, London was equipped with a network of steam-powered underground railway cars, and the first electric railway was installed.

The 20th Century

> "All human progress, political, moral, or intellectual, is
> inseparable from material progression."
> —Auguste Comte

Never in history has humankind displayed such a collaborative, concerted and pragmatic spirit to go beyond existing limitations. In the beginning of the book, I compared the determination and resilience of humankind with that of a boxer whose efforts are only encouraged by each knockdown. Let's look at a few facts.

This was the century of the devastation caused by the Spanish flu, World Wars I and II, and the Great Depression of 1929. Other major political events include the Cuban Missile Crisis; the wars in Korea, Vietnam, and the Middle East; the petroleum crisis; the emergence of international terrorism; and unspeakable genocides. The list could go on and on.

However, on a positive note, there were many incredible advancements: the continuation of the widespread combination of technology, capital, and labor, and the scientific management of production of Frederic Taylor in the manufacturing process started in the preceding century gave people the sweet taste of health and wealth. Penicillin and a vaccine against tuberculosis were discovered in the first half of the century; during that same period, polio was about to be eradicated. By 1979, smallpox was eliminated based on the original work of Edward Jenner, the patriarch of mass vaccination. The number of millionaires grew from 100 in 1870 to 16,000 in the first decade of the 20th century. The rebuilding and stabilization of Europe through the Marshall Plan after World War II and the end of the Cold War are positive achievements that have given hope for a better future.

Technology has become a reliable ally, assisting in the improvements of production efficiency and providing the free world with protection, over and above expectations of living conditions in terms of accommodation, communication, medicine, and entertainment. The hegemony of the United States has been well established. Hundreds of inventions and innovations took place in the 20th century, covering various areas of human activity. More important, however, are the far-reaching ramifications of some of those inventions, particularly in the high-technology industry, medicine, transportation, and communication. They reflect the evolutionary progress of the inventions, innovations, and discoveries of the preceding centuries. Cars, computers, and the internet are vibrant examples of this phenomenon.

Many theories put forward in earlier periods have been either revisited from a different perspective, regardless of how sound they may have appeared, or have been repealed because of new scientific evidence. Bryan Greene, one of the world's leading physicists, in the preface of his bestselling book *The Fabric of the Cosmos* (p. 8) outlines the dynamic, evolving nature of science with the passing of time. His comments fit well in this context. He mentions that questions answered by one generation are overturned by their successors and refined and reinterpreted by scientists in the centuries that follow. For example, on the issue as to whether empty space is a real entity or an abstract idea, Isaac Newton in the 17th century declared that it is real. Ernst Mach in the 19th century concluded that it isn't. Albert Einstein in the 20th century reformulated the question by adding the notion of space and time and refuted Mach. Further space-based experiments confirm particular features of Mach's conclusions that happen to agree with Einstein's general relativity theory. Basically, science is and will continue to be in the making. It seems that the philosopher of science Karl Popper has the last word. A theory is falsifiable if there is an "inherent possibility that it can be proven false. Science progresses by the successive rejections of falsified theories. They are replaced by newer theories providing greater explanatory power." Popper claimed that if a theory is falsifiable, then it is scientific.

The 20th century introduced technological inventions that we still do not fully understand and whose ramifications and social consequences cannot yet be assessed. If Johannes Gutenberg revolutionized the dissemination of information with the movable type printing press, the internet has globally unified and made accessible the world's knowledge in an effective and efficient vehicle affecting the socioeconomic and political playing fields. Humankind has entered the information age, and a unified bank of knowledge about everything is within our reach and in a timely manner, regardless of our geographical constraints. We are living in a virtual world in addition to the real world.

Some of the most important inventions and discoveries of the 20th century were:

- Radio   ~ 1901 by G. Marconi
- Automobile   ~ 1908 by K. Benz 1890 & H. Ford
- Insulin   ~ 1921 by F.G. Banting & C.H. Best
- Liquid fuel rockets   ~ 1926 by R.H. Goddard
- Television   ~ 1928 by P. Farnsworth
- Penicillin   ~ 1928 by A. Fleming
- Programmable computer   ~ 1941 by K. Zuse
- Needle-free injector   ~ 1942 by Robert A. Hingston
- Transistor   ~ 1947 by Shockley, Brattain & Bardeen
- Vaccine (polio)   ~ 1952 by Jonas Salk
- Stem cell   ~ 1963 by J.E. Till & E. McCulloch
- Artificial heart   ~ 1982 by Robert Jarvik
- Mobile phone   ~ 1970 by Martin Cooper
- Internet   ~ 1983 by ARPANET
- 3D printing   ~ 1983 by Chuck Hull
- World Wide Web (www)   ~ 1991 by Tim Berners-Lee

It was quite a challenge to come up with the most "important" inventions of the 20th century, considering the immensity of areas covered by this expanded technological explosion, which was to continue into the next century. In this context, importance as a selection criterion means relevance and usefulness to society. As Isaac Asimov put it in another context: "One can judge scientific advance by its effect on the world today....Discovery and inventions can only count when they affect society." I believe that the above list meets this criterion.

Again, all of the above inventions are important within the meaning of the aforementioned definition. However, I believe that I express a general consensus in saying that at least three of the above listed innovations—namely, the needle-free injector, the transistor, and the internet—stand out because of their originality, effectiveness, and ramifications to the present day and in the years to come. The needle-free injector since its invention has been perfected to deliver drugs and vaccines. It is in this latter use that this great invention will benefit society, particularly in developing countries. I hope that there will come a time when blood samples can be taken without a needle. The author of this book is horrified of needles!

The transistor, a tiny piece of silicon and trace elements of other materials, became the major component of integrated circuits, revolutionizing the electronic industry in terms of the price–performance ratio of radios, televisions, cars, home appliances, and telephones, and that is just the beginning. The internet shattered geographical limitations, and the way we live has substantially changed and will continue to change with the advent of social media. Welcome to the virtual world!

With the development of the transistor and the integrated circuit, the last decade of the 20th century saw an unprecedented growth of inventions in the field of computers. The development of information technology brought the relationship between humanity and machines to a new level that would continue into the next millennium. Technology as an ally for the pursuit of a common objective continues its course. It is a relationship in which the destiny of both humans and technology is more intertwined. Technology now makes up, perceptibly, a greater part of our life, and it can be taken for granted that this trend will continue unimpeded into the future.

Technology is indeed unpredictable. For example, nobody could have foreseen the evolution of cell phones into smaller and smaller sizes or the beginning of social media in the last decade of the preceding century and the first touchscreen phone by IBM in 1992. Even certain expressions are outdated. We do not dial a phone number anymore. In the same vein, we do not use the term color TV. We just expect a TV to provide color pictures, relegating the black and white monitor to specific technical work or to the recycling bin. It is a cyclical process in which one invention is improved until it reaches a tipping point calling for a new and different technology to respond to our insatiable needs. At best, one can call this fate evolutionary, or at worst planned obsolescence, compounding our current environmental challenges.

But the review of the inventions and discoveries timetable reveals another phenomenon of paramount importance: the occurrence of a similar pattern with increasing momentum over time, the quality of human life on Earth, the obvious and the ever constant will to explore the cosmos, the inner call of humanity. Pioneer 10 and 11 and Voyageur 1 and 2 for the exploration of the outer solar system are vibrant examples. More significant, however, was the technological development process leading up to the moon landing in 1969. Data about Jupiter were successfully transmitted, and it is now believed, because of the twin Mars Exploration Rover landed in the early 2000s, that Mars may be habitable. Our old cliché about Mars has been shattered. As in the case of the moon exploration, certain inventions and discoveries had to be in place to make these explorations possible.

Looking back at all those inventions and innovations, the case can easily be made that life has improved. In all developed countries, a fridge, stove, television, and telephone are common, and one would be hard pressed to find a household without at least one of these items. No one would think of turning the clock back.

The 21st Century

We all should have felt good about the new millennium. Contrary to the naysayers, humanity has smoothly moved into a new century, defying the Y2K myth. Regardless of a series of natural disasters (earthquakes, mudslides, tsunami, and hurricanes) and human-made mishaps (nuclear plant accidents, terrorist attacks), we have witnessed the blooming of some of the theories put forward in the previous century and the emergence of new technologies, such as biotechnology and nanotechnology.

Technology continues to evolve towards greater complexity and sophistication while maintaining an appearance of simplicity. In the first decade of this century, we see the extension of information technology, computer sciences, and nanotechnology, and the continued contribution of the merger of disciplines in many areas, particularly in medicine, agriculture, and engineering, let alone the creation of the USB flash drive (2000), Mozilla Firefox (2004), YouTube (2005), smartphones (2007), and a whole range of computer games. But apart from these novelties in communication and information technology, there is a much deeper sense of the beginning of a new era. For instance, at the time of writing, Google announced the possibility of putting driverless cars on the road by 2019. This will be a profound change with deep social, ecological, and economic ramifications; fewer bottlenecks on the highway, fewer accidents, less pollution, fewer unproductive hours missed because of heavy traffic. In the USA, these costs have been estimated to total more than $150 billion a year. In Canada, the estimated annual cost of congestion in Toronto could be up to $11 billion yearly. This latter figure is part of a study conducted by the C.D. Howe Institute and released in July 2013 (12). According to a report of the Greater Toronto-Hamilton Area, (13), (14) "every hour spent in a car on a daily basis is associated with a 6% increase in the likelihood of obesity." This latter in turn increases the risk of heart disease, diabetes and cancer. As mentioned in the report, improvements to road infrastructure can alleviate the problem. But in the same vein, I also believe that the driverless car contains the means to address this challenge through better traffic management, ride sharing, and elimination of stress for drivers. By the next decade, the driverless car will change the entire transportation system, including car ownership itself.

Among the most important inventions and innovations of the 21st century to 2010 are:

- Smartphone   ~ 2000
- iPod and iTunes   ~ 2001
- Electric car   ~ 2003
- Facebook   ~ 2004
- YouTube   ~ 2005
- iPhone   ~ 2007

A longer list is provided in Appendix II. Again, I do not pretend that this list is all inclusive, given the discrepancies between the various sources of information, but it does show the extent of technology in all areas of human affairs. In the words of Steve Van Dulken ("Inventing The 21st Century" blog), "From the iPod to the Nintendo Wii, the first decade of the 21st century has already brought us incredible inventions we could not have imagined and that have already changed how we live and spend our free time." Technology is changing more rapidly than ever. The explosion of social media seems to have brought people closer. This virtual bond is made at the expense of in-person interaction. Without over-emphasizing the importance of physical attraction, it has become possible that even a higher feeling such as love may occur between people who have never physically met each other. The connectivity created by the digital revolution of the 20th century has led the Fourth Industrial Revolution (4IR). Professor Klaus Schwab, Founder and Executive Chairman of the World Economic Forum describes this revolution in these terms:

> "The First Industrial Revolution used steam power to mechanize production. The Second used electric power to create mass production. The Third used electronics and information technology to automate production. Now a Fourth Industrial Revolution is building on the Third. It is characterized by a fusion of technologies that is blurring the lines between the physical, digital, and biological spheres."

Indeed, there is an ongoing fusion between man and machines, as can be seen in electronic wearables, customized prosthetics, and more sophisticated modes of communication. Manufacturing will become more and more automated, and robotics and artificial intelligence will play a more preponderant role in our lives. Adaptation to this new way of life is slowly taking place as people start enjoying the benefits of these new technologies. But this revolution will also come with a social cost if the education system does not equip people with the proper skills to take advantage of the new opportunities available.

On a broader basis and in view of the multifaceted aspects of technology since the Renaissance, could technology be considered a phenomenon falling in one of the following categories?

- The spirit of the time (zeitgeist)
- Perpetual renewal from age to age
- Continuing improvement

A realistic answer is all of the above. Technology can be seen as a dynamic entity in perpetual metamorphosis. As a materialization of ideas, technology in the form of innovations reflects the spirit of the time (zeitgeist), such as in the surge of renewable energy initiatives leading to the spread of solar panels since the energy crisis of the 1970s. In the same vein, electric cars are becoming more and more of a reality in the current century. However, technology is not constrained by zeitgeist, because of its creative and proactive nature. Technology has renewed itself to limitless capacity from the mechanical age of the First and Second Industrial Revolution to the automation age. Since the Digital Age, the Third Industrial Revolution starting in the last decades of the preceding century, the accelerating pace of innovation improvements and the combination of academic disciplines have led to a Fourth Industrial Revolution (9, 10), more profound than the preceding revolutions. A common theme between the Third and the ongoing Fourth Industrial Revolution is a trend towards less reliance on fossil fuel and an increased connectivity because of the sophistication of information technology. The manufacturing process will be more automated and will take the production of goods to an unprecedented level of quality and precision. In general, technology moves from slow change to rapid, sporadic, scattered, and unrecognizable results—or, in the words of Kevin Kelly, among other attributes, from simplicity to complexity. This latter phenomenon is not new. Since the time of Aristotle, this complex and unexplainable metamorphosis in technological development to meet human needs has caught the attention of many philosophers. The complexity theory of the famous mathematician and physicist Stephen Wolfram states that this complexity theory can shed light on the nature of technology. The theory "seeks explanation for apparently unpredictable phenomena…in the interplay of their myriad simple components" (11). The consultant Marcus Jenal in his blog "What Is Complexity?" referring to an article of David J. Snowden and Mary E. Boon published in the November 2007 issue of the *Harvard Business Review*, identified the following aspects of complexity:

- It [a complex system] involves large numbers of interacting elements.
- The interactions are nonlinear, and minor changes can produce disproportionately major consequences.
- The system is dynamic, the whole is greater than the sum of its parts, and solutions can't be imposed; rather, they arise from the circumstances. This is frequently referred to as emergence.
- The system has a history, and the past is integrated with the present; the elements evolve with one another and with the environment; and evolution is irreversible.
- Though a complex system may, in retrospect, appear to be ordered and predictable, hindsight does not lead to foresight, because the external conditions and systems constantly change.
- Unlike in ordered systems (where the system constrains the agents), or chaotic systems (where there are no constraints), in a complex system the agents and the system constrain one another, especially over time. This means that we cannot forecast or predict what will happen.

Now we can appreciate why predictions about technology and, for that matter, the future of humanity will remain difficult if not impossible. Too many moving parts! However, this is not to say that trends cannot be plotted.

The following graph displays the progression of technology since 1450 to the first decade of the current millennium. As can be seen, inventions and innovations have occurred at an accelerating pace since the 19th century—a reflection of our needs and wants. The diffusion of information by the Gutenberg movable type printing press, followed by the publication of the *Encyclopedia* of Diderot and, more recently, computers, television, and the internet, has substantially contributed to this growth. Among the most important inventions in the recent past are electricity in 1882; the first (mechanical) computer in 1837; and, in the 20th century, the transistor.

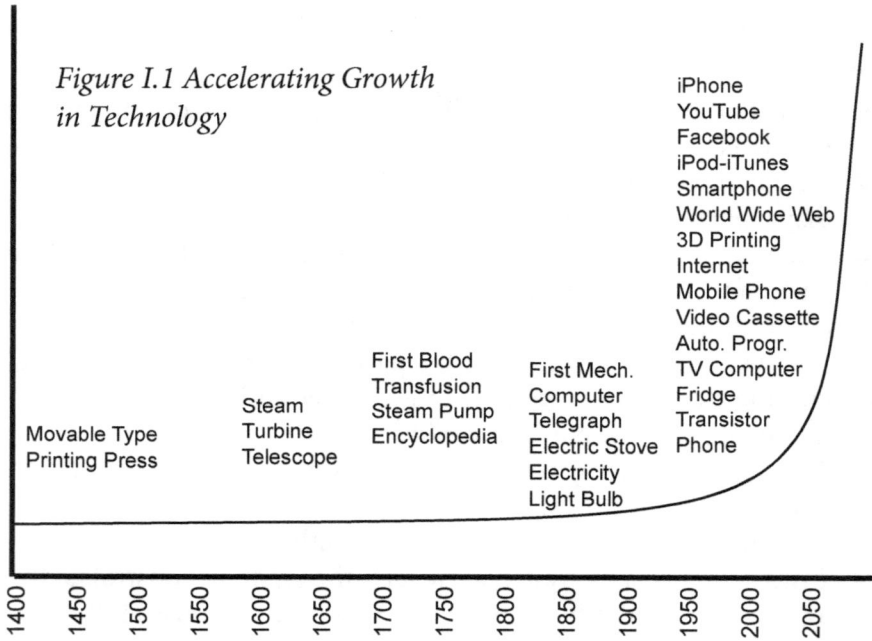

Figure I.1 Accelerating Growth in Technology

	iPhone
	YouTube
	Facebook
	iPod-iTunes
	Smartphone
	World Wide Web
	3D Printing
	Internet
	Mobile Phone
	Video Cassette
	Auto. Progr.

| Movable Type Printing Press | Steam Turbine Telescope | First Blood Transfusion Steam Pump Encyclopedia | First Mech. Computer Telegraph Electric Stove Electricity Light Bulb | TV Computer Fridge Transistor Phone |

1400 1450 1500 1550 1600 1650 1700 1750 1800 1850 1900 1950 2000 2050

Suggested Readings

(1) Isaac Asimov, *Chronology of Science and Discovery*. New York: Harper and Row Publishers, 1989.

(2) "Technological Advances: Discovery, Invention, Innovation, Diffusion, Research and Development" http://thismatter.com/economics/technological-advances.htm

(3) Jared Diamond, *Guns, Germs, and Steel: The Fates of Human Societies* https://www.amazon.ca/Guns-Germs-Steel-Jared-Diamond/ dp/0393317552

(4) "A Brief Excursion into Three Agricultural Revolutions"
http://climate.umn.edu/doc/journal/kuehnast_lecture/l4-txt.htm

(5) Daily Life in the 17th Century in England
http://www.localhistories.org/stuart.html

(6) The Revolutionary Babbage Engine. Unprecedented. Unparalleled. Unfinished
http://www.computerhistory.org/revolution/calculators/1/51

(7) Letter to Madame Christina of Lorraine, Grand Duchess of Tuscany
http://inters.org/Galilei-Madame-Christina-Lorraine.

(8) Brian Green, *The Fabric of the Cosmos: Space, Time and the Texture of Reality*. New York: Vintage Books, 2004.

(9) Industrial Revolutions
www.sepaforcorporates.com/thoughts/heck-fourth-industrial-revolution/

(10) The Fourth Industrial Revolution
https://www.weforum.org/agenda/archive/fourth-industrial-revolution/

(11) "The Man Who Cracked the Code of Everything"
www.wired.com/2002/06/wolfram/

(12) Toronto Gridlock May Cost the Economy up to $11 Billion Annually
http://www.cbc.ca/news/business/toronto-gridlock-may-cost-economy-up-to-11b-c-d-howe-says-1.1394574

(13) Improving Health by Design
https://www.peelregion.ca/resources/healthbydesign/pdf/moh-report.pdf

(14) Franl LD, Andresen MA, Schmid TL, Obesity relationships with community design, physical activity. and time spent in cars. Am J Prev Med 2004; 27(@):87 - 96

CHAPTER II

An Inevitability

"I believe in intuition and inspiration. Imagination is more important than knowledge. For knowledge is limited whereas imagination embraces the entire world, stimulating progress, giving birth to evolution. It is, strictly speaking, a real factor in scientific research."
—Albert Einstein

"Civilization advances by extending the number of operations we can perform without thinking of them."
—Alfred North Whitehead

The sedentary society that existed 12,000 BP is referred to as the cradle of civilization. This new mode of living, with rudimentary farming, building, clothing, and cooking tools, evolved into a refined lifestyle with sophisticated agricultural implements, as well as extensive manufacturing, communication, and transportation modes. From a utopian perspective, we are on the right path. We should then describe our observation of this growth, because what we did and are doing now in this accelerating pace of technological development will redefine us as a society. However, from a dystopian perspective, we may be on a path leading to domination by our own creations if we do not apply enough wisdom when developing certain of these innovations.

Originally, I had entitled this this chapter "A Path to Complexity," but on second thought, I decided that "Inevitability" would be more appropriate in the context of technological development because the former leads to the latter. The universe works from order to disorder; in a complex system, each stage of creation contains the seeds of other creations, provides the information for further diversification to occur when the proper conditions are met. Big History provides impartial responses to big questions such as: What is life? How did iron make it to our planet? How did the Earth form? The second law of thermodynamics explains the development of complexity.

David Christian, in a video entitled "The History of our World in Eighteen Minutes" (4), calls these threshold stages goldilocks or the proper circumstances for the evolution of our planet and humanity. Some goldilocks have greater impact than others. In our species, just consider the advent of language some 50,000 years ago, the transformation from hunting and gathering to the formation of sedentary societies, and the invention of the movable type printing press more than 575 years ago. These three events have a common denominator: the transmission of information accelerated by our collective learning. The key mechanism has always been and will always be the transfer of information from one generation to the next.

My observation of technological development over the last few centuries suggests that it is:

1. Independent of economic movement.

2. Fed by its own internal momentum through the combined effect of creativity and scientific development.

3. Unstoppable and therefore cannot be suppressed.

1. Technological Development Is Independent of Economic Movement

According to the U.S. National Bureau of Economic Research (NBER), there have been only two economic depressions: the Long Depression of 1873–1879, and the Great Depression of 1929–1933. However, there have been as many as 47 recessions in the United States since 1790. Among them are those of 1815, 1828, 1920, 1937, 1953, 1973, and 2007. This last, also called the Great Recession, ended in 2009. A depression is more severe than a recession. A common joke in economic circles may serve to illustrate the difference between the two. When your neighbor loses his/her job, this is a recession. When you lose your job, that's a depression. I suggest that technological development progresses irrespective of economic depression or recession.

Depression

A depression is a sustained downturn in the economic activity of one or more economies, lasting two or more years. The Long Depression (LD), resulting from the collapse of the Vienna Stock Exchange and fueled subsequently by the bankruptcy of American banks, started in 1873 and ended in 1879, lasting approximately six years in the USA and far longer in some European countries. In order to test the neutrality of technological development with respect to economic movements, I have carried out a semi-rigorous analytical review of the total number of inventions, innovations, discoveries, and other scientific observations made during the six-year period preceding the LD (1866–1872) and the six-year period of the LD (1873–1879). I have used the *six-year* duration of the LD as the timeframe for the comparison. I qualify this review as semi-rigorous because of the constraints imposed by the database. The review did not show a diminution in the number of inventions, discoveries, and other scientific observations during the LD. Moreover, some important discoveries, such as platelets (a colorless cell fragment without a nucleus, found in blood) and gallium (a soft, silvery-white material used in semiconductors) occurred during the LD. Similarly, inventions such as the radiometer and phonograph occurred during the LD.

1866–1872: 26 inventions, discoveries and other scientific observations

1873–1879: 29 inventions, discoveries and other scientific observations

The Great Depression (GD) of 1929–1933 was caused by a multitude of factors and resulted in the U.S. stock market crash of October 29, 1929. This crash is also known as Black Tuesday. Causal factors, including unregulated banks, inappropriate monetary policies, under-consumption of goods and services, and tariff barriers, have been the subject of long debates between economists and politicians but are outside of the scope of this book. The GD ended in 1933. Using the approach described above, I reviewed the total number of inventions, innovations, discoveries, and other scientific observations during the *four-year* period preceding the GD (1924–1928) and the four-year period of the GD (1929–1933). As above, no diminution was found during the review period in the number of inventions and discoveries. Moreover, it was during the GD that the planet Pluto was discovered, and nylon and Polaroid were invented.

1924–1928: 37 inventions, discoveries, and other scientific observations

1929–1933: 48 inventions, discoveries, and other scientific observations

Recession

According to the NBER, recession can be defined as a significant decline in economic activity characterized by a decrease in the real gross domestic product over two consecutive quarters and lasting more than a few months. As noted above, there have been as many as 47 recessions in the USA since 1790. I must add that the NBER, founded in 1929, does not date recessions prior to 1857. Economists extrapolate the number of recessions of the preceding period based on certain known major events associated with recession, such as war and banking crises. In addition, the duration of a recession, typically less than a year, makes the consistency of using one source of information more difficult. The above-named constraints have resulted in a limitation of the database. The review of a sample of 12 recessions from 1887 to 2000 revealed that in 70% of the recessions, the number of inventions, discoveries, and other scientific observations exceeded the number in the pre-recession period.

On the basis of this semi-rigorous calculation, it appears that inventions and discoveries progress irrespective of economic depression or recession, but innovations do not, as will be discussed in the next section.

2. Technological Development Feeds Its Own Internal Engine

As discussed in the preceding chapter, from the 19th to the 21st centuries there has been an acceleration of technological development. Each invention attracts related or disparate inventions, as evidenced by the mosaic of inventions in the information technology sector that have overflowed into the manufacturing sector. At this point, it is important to differentiate between the role of market forces in innovation compared with their role in invention. What communication lines or contacts exist between the inventor and the consumer? A short answer is: none. However, there are always exceptions. A few of them are the invention of the cotton gin, the steam engine and, in the not too distant past, the atomic bomb because of the necessity of the times. Inventors are self-motivated by the urge to extend the boundaries in their particular field. Market forces drive innovations, not inventions. Innovations derive from inventions for profit in the free-market system, the objective of most entrepreneurs. They must excel at detecting the needs and wants of potential consumers, and to a certain extent the affordability of their products or services for those consumers. They capitalize on opportunities to grow their businesses. Needs and wants drive innovations; therefore, because of reduced consumer spending levels during a recession, innovations are negatively affected.

Certain discoveries and inventions may occur from pure luck or accident in the trial process, such as the development of penicillin (8B) and rubber (8A), or, as mentioned earlier, a phenomenon called simultaneous invention, as in the case of oxygen (8D). Furthermore, for market opportunity reasons, not all inventions make it to the market in the form of innovations, depending on their usefulness and/or desirability. In fact, many inventions have been transformed into quite different items than what had originally been invented. One example is the stent. Originally invented in the 19th century by dentist Charles Thomas Stent, an organic compound used as a mold for dental impression has later been used in plastic surgery and, since 1986, mostly in cardiology as a mitral valve replacement (8C). In the same vein, mobile phones should be seen as the platform for tablets and video players, which are examples of incremental innovations. These ideas come at the right time and are propelled and created because of the market needs.

From a macro-economic viewpoint, both inventions and innovations are important for the economic development of a country. Expenditures on innovations by both firms and consumers increase the growth phase of an economy and have a positive impact on the gross domestic product (GDP). In addition, inventions and innovations drive a company's competitive advantage in national and international markets. The value of a company is based not only on its business volume in the marketplace but also on the number of new products and services yet to be commercialized. Ooops! I have digressed from the subject of this chapter!

3. Technological Development Is Unstoppable

Technological development cannot be vanquished. Like gravity, it can only be artificially suppressed. Observation (c) becomes almost a corollary of observations (a) and (b). It is like a machine gaining increasing momentum over time because of constant acceleration. In one of his bestselling books, *The Singularity* Is Near: When Humans Transcend Biology* (19), Ray Kurzweil demonstrated that, on the basis of data gathered so far, technological growth doubles exponentially in that not only the growth but also the rate of growth itself is growing exponentially, as per the Law of Accelerating Returns (LOAR). This leads to the fact that the unit cost of an item fabricated at a certain time is a function of the workers' skills, the state of the technology, and the source materials.

The improved price-performance ratio in the high-tech industry is the materialization of Moore's Law. Based on observations of the history of computing hardware, it states that the number of transistors in a dense integrated circuit doubles every two years. Reinforced by the fact that the high-tech industry has used this prediction to set production targets, Moore's Law has been concretized for the past 40 years. Perhaps in a self-fulfilling prophecy, all electronic devices have been improving at a growing rate. Better performance with increased miniaturization at a lower price is the trend for the foreseeable future of all electronic devices, from the briefcase mobile phone model to the pocket-size smartphone. Moore's Law can be seen as the driving force behind most technological changes and the resulting productivity and economic growth of the late 20th and early 21st centuries in the electronic industry. However, since this law is not a physical or natural law, like gravity or the laws of thermodynamics, there is no ground for extrapolating that this historical growth will continue indefinitely, unless Moore's is replaced by a more promising law. In this eventuality, the impact of technology on the labor market should not be understood as a short-term issue.

When I was young, I was fascinated by multicolor kaleidoscopes. At that time, most of them were produced as a cylindrical tube with inlaid glass of different colors. When the light passed through the tube, the magnificence, variety, and order of the images projected

---

* For the sake of simplicity, I have retained, among others, this definition of singularity by James Martin, a world-renowned futurist, computer scientist, author, and lecturer: singularity "is a break in human evolution that will be caused by the staggering speed of technological evolution."

kept me focused and serene for a long time. I must confess that even today, watching the orderly sequences of geometric forms triggered by a song played on my laptop is one of my hobbies. An analogy can be made between the kaleidoscope and technology. Both display infinite output within a certain time/framework, multiple mixtures of "forms," and an infinite number of creations. Both are initiated by humans and have the singular capacity to continue forever. In the multidisciplinary approach to address the challenges of our time, technological development is unlimited and unpredictable in all fields of application. Another analogy would be the infinite expression of ideas through the arrangement of letters or characters in any language.

We are witnessing a transformative process in humankind's thinking and activities, promising limitless capabilities as part of technological development. We are making machines that we cannot even see. Eric Drexler, in his book *The Engines of Creation: The Coming Era of Nanotechnology*, wrote that our ability to arrange atoms lies at the foundation of nanotechnology. The operating scale of nanotechnology can be demonstrated by analogy with the thickness of a sheet of paper, which is 100,000 nanometers thick, or a human hair, which is approximately 80,000 to 100,000 nanometers wide. This new technology will manage individual atoms and molecules with control and precision; in nanotechnology, molecules are measured in nanometers (10^{-9} of a meter). The things around us act as they do because of the way their molecules behave, and they behave according to how the molecules are arranged. The performance of human proteins is based on preset functions. More complex patterns make up the active nano-machines of living cells. As the microbiologist Louis Pasteur wrote a couple of centuries ago, "The role of the infinitely small is infinitely large." It becomes clear that the flow of current events, their acceleration and direction, will change medicine, engineering, and robotics in a way that we could not have imagined. This is quite an achievement for a species that, based on a cosmic time scale (12), has only recently began walking upright when it moved from forest to grassland. What factors have prompted such a development? How did we get there?

Well, as part of the evolutionary trend and since the beginning of human life on Earth, once we learned how to start and maintain a fire, and once we created the wheel, we put into motion an unstoppable engine of creating products and mechanisms to protect ourselves and improve our lives. This engine has grown in complexity over time, and has introduced and will continue to generate many tangible benefits through a succession of technological and social changes affecting us. I cannot think of a better way to describe this phenomenon than by quoting Marshall McLuhan, who wrote in *The Gutenberg Galaxy* that "technologies are not simply inventions which people employ but are the means by which people are re-invented." Indeed. For better or worse, we, the recipients of technological novelties, have mentally changed by relying more on the knowledge and control of the universe around us than on believing in the effects of spiritual forces.

A picture is worth a thousand words. The following pictures of some inventions, from the time of their creation to our current epoch, are intended to facilitate an appreciation of technological evolution over time, including the socioeconomic ramifications.

The Spinning Wheel

The invention of the spinning wheel lifted a heavy burden, that of twisting fiber into yarn. A foot-powered treadle turned a large wheel, which twisted a spindle instead of this having to be done manually. This improvement increased the productivity of the family unit, considering that this work was primarily carried out at home, mostly by women.

As time went by, the desire of merchants to increase production, combined with greater efficiency in the agricultural techniques of cotton production, moved such artisanal-type industry from the home to a factory conceived for this purpose. As one can imagine, a factory is an environment more conducive to productivity.

Urbanization, increased capital investment in machinery, and the emergence of a concentrated labor class were some of the ramifications of this technological development.

Figure II.1 The spinning wheel

The Mechanical Clock (1300)

The attainment of a desired production level would not be possible without the measurement of time. In a factory environment, when remuneration is on a piece-work basis, a worker can produce enough to meet his/her family's needs, but this might not necessarily meet the owner's business goals. This voluntary slowdown was called "leisure preference" by S. Marglin in her article "What Do

Bosses Do? The Origins and Functions of Hierarchy in Capitalist Production," published in the *Review of Radical Political Economics*. The clock serves as an activity monitor, a performance-control instrument subsequently used by F.W. Taylor in time-motion studies. One can anticipate the laboring class's non-receptivity toward this action even today, particularly those doing piece work in the textile industry. George Woodcock (5), a Canadian writer, in the The *Tyranny of the Clock* exposed the consequences of this invention, which became popular in the 14th century. Time has become a commodity that can be sliced, compressed, and sold by business owners for their own gain.

The Movable Type Printing Press (1450)

Figure II.3 illustrates the spread of the movable type printing press throughout Europe, making possible the dissemination of scientific and literary information. This map was taken from an insightful article entitled "Information Technology and Economic Change: The Impact of the Printing Press" (6), written by J. Dittmar, PhD.

Figure II.2 The printing press
16th century – art and craft print workshop illustration

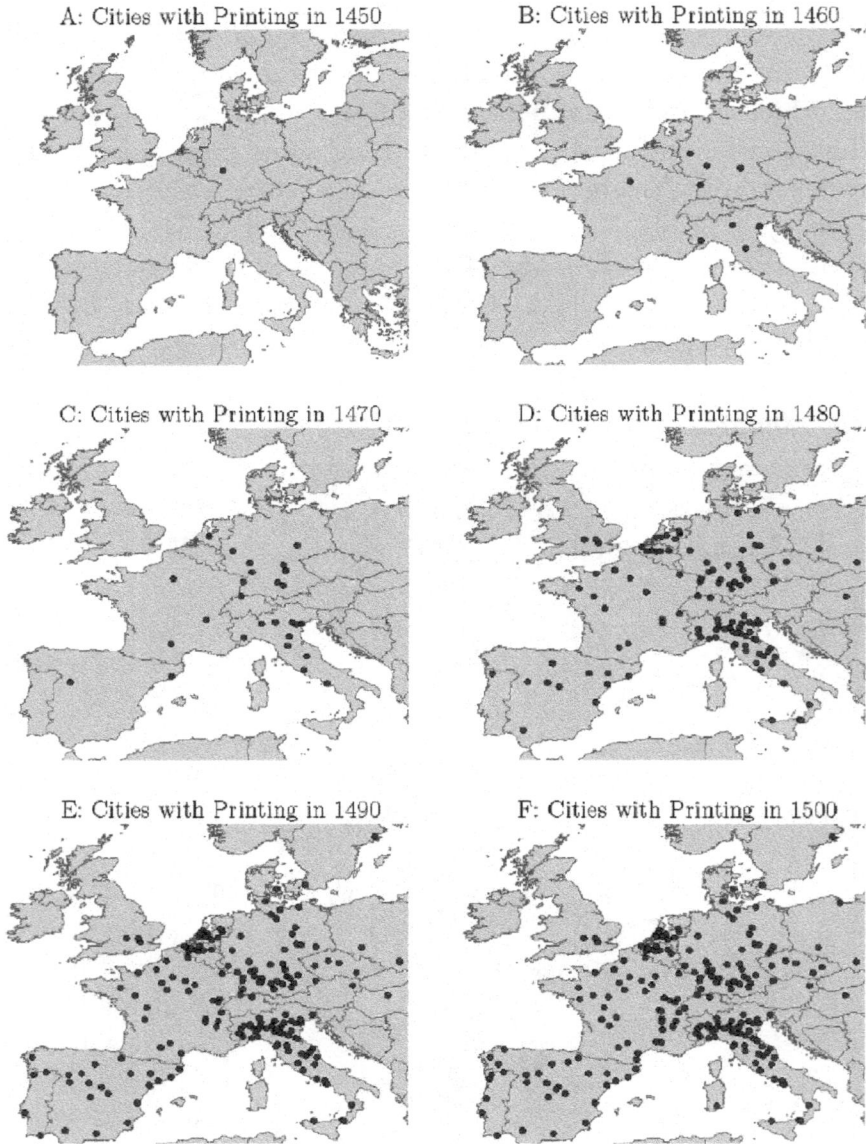

The Diffusion of the Movable Type Printing Press

A: Cities with Printing in 1450

B: Cities with Printing in 1460

C: Cities with Printing in 1470

D: Cities with Printing in 1480

E: Cities with Printing in 1490

F: Cities with Printing in 1500

Figure II.3 The diffusion of the Gutenberg movable type printing press
Source: Reproduced with the permission of J. Dittmar, PhD

In this article, he outlined three factors contributing to cities' development because of the printing press:

- The printing press was an urban technology, producing material for urban consumers.
- Cities were seedbeds for economic ideas and social groups that drove the emergence of modern growth.
- City sizes were historically important indicators of economic prosperity.

It is a fact that the agglomeration of people facilitates collective learning, and the printing press gave momentum to this phenomenon. One can appreciate this statement by considering the urbanization trend in the recent past. Cities have become centers of attraction to satisfy cultural and socioeconomic needs.

Although the printing press benefited society in general, it should also be noted that not everyone was literate in the 15th and 16th centuries. Similarly, today not everyone is literate or has access to a computer. However, as time went on, the technological revolution spurred by the printing press sustained the cultural and the scientific revolution of the Enlightenment period.

The Steam Engine (1781)

To be accurate, the steam engine had been operational for almost one hundred years earlier than 1781 and was used in factory and coal-mining operations. However, the modifications made by James Watt in 1781 rendered it more efficient. As a result, the steam engine was subsequently put to use not only in factories but also in transportation, especially railway locomotives, marine voyages, and mining.

In the words of Isaac Asimov in *The Chronology of Science and Discovery*, "The steam engine, bringing the use of energy to all mechanical devices in far greater quantity than anything else had to offer in the past, was the key to all that followed rapidly under the name of the Industrial Revolution, when the face of the world was changed so drastically and far more rapidly than at any time since the invention of agriculture, nearly ten thousand years before."

The steam engine has remained one of the foundational inventions for industrial development.

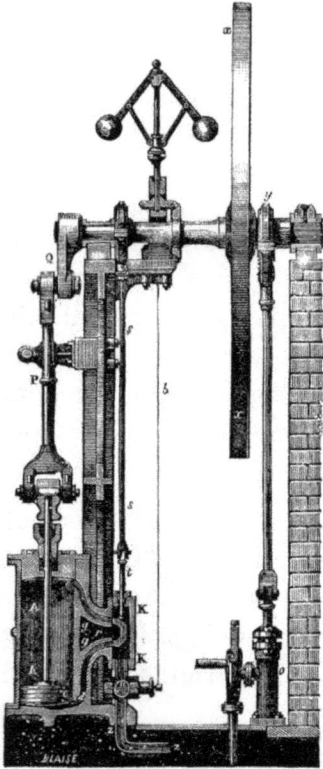

Figure II.4 The steam power machine and associated devices

Emerging Technology

The preceding and succeeding decades of the new millennium witnessed impressive steps in technological achievements. The perfection of computers, the explosion of the internet, the development of information technology, the interdisciplinarity of sciences within biotechnology, and nanotechnology have expanded the horizon of human capability. Agriculture, communication, manufacturing processes, and health are being reformed in a singular way. For instance, our knowledge about the functioning of the brain raises the hope of finding cures for schizophrenia, Alzheimer's, and autism. But this knowledge will also benefit the development of artificial intelligence (AI) and many industrial and commercial applications. We will have to face the dilemma of whether might makes right. In other words, we may or we may not want to build some "stuff" simply because we can. The associated risks or ethical considerations must be taken into account. This important issue will be addressed later.

All of these activities can be seen as parts in a unified force to conquer the future, to find the magic formula that will eliminate all our technical limitations. With nanotechnology, matter seems finally to have been conquered because we now have access to its very essence, its building blocks at the nanoscale level. Over time, everything will be built more economically than we have ever seen, bringing society to another level. Never before has *Homo sapiens* been able to assemble such a powerful and unlimited force. In the foreword of K. Eric Drexler's book *Engines of Creation,* Professor Marvin Minsky refers to the numerous ways to stack atoms when he writes, "What we can do depends on what we can build." It would seem then, theoretically speaking, that since atoms can be rearranged in myriad ways, we are at the dawn of a new era. The future benefits of nanotechnology in the health sector are covered in the next chapter. They are limitless. A few benefits of genetics and robotics are already tangible. With the progress made in the field of DNA sequencing, SARS (severe acute respiratory syndrome) has been cured, and AIDS (acquired immune deficiency syndrome) is better managed; there is reason to hope that a cure will be found in the next decade. A whole range of treatments is being studied using gene transfer between somatic or germ cells. While in the former the resulting effect will not be passed to future generations, an error may lead it to the latter. The technology for germline intervention is not secure enough at this time to warrant its use in humans (16). However, there may come a time when the burden of incurable diseases could eventually force humankind to reconsider certain disruptive nanotechnological and genetic engineering procedures.

We have seen strong improvements in the contraceptive field. The OECD report entitled *21st Century Technologies – Promise and Perils* states: "The contraceptive technology developed in the past century has separated procreation from recreation." As a mature society, we have come to the realization that love expression may not have to necessarily lead to procreation. In terms of our emotional quality of life, we now have the choice to respond to our biological impulses. Smart pills and a variety of contraceptive methods for both men and women are available. Nevertheless, contraceptive pills are still not available in most developed countries without a prescription, perhaps for health security reason. In addition, some countries still do not support the use of protective sex wear. Society does not always keep pace with technological advances.

In the area of medical assistive technology, limb and organs replacements are a necessity, especially for injured soldiers returning home and deserving a normal life. Prosthetics are no longer rigid limb structures but have evolved into esthetic replacements with sensor implants, allowing the reproduction of gestures and activities by a particular limb.

But it is in the capabilities arising from the interdisciplinarity of the sciences that great inroads into eradicating the top three killer diseases of heart failure, cancer, and diabetes will be made. Coupled with the rapid development of information technology, early diagnosis, monitoring, and therapeutic treatment of these diseases will be far more effective. Heart failure is not only a growing and costly problem for our society but also one associated with high morbidity and mortality. It is estimated that in 2013, 56 million people died of heart disease around the world (49,000 in Canada and 900,000 in the United States). Tremendous progress has been achieved so far in the implantation of artificial devices to resynchronize the diastolic and systolic pressure of the heart. In the not-too-distant future, it will even be possible to replace the heart altogether if replacement is the best solution in a specific case. This strategy is conditional upon further developments in synthetic biology and the mastering of the brain functions. Internal and external organ replacement will remove mobility limitations, and more efficient medications will have a positive effect on our living conditions. In the manufacturing sector, production is becoming more and more automated. Cars, medical drugs, textiles, and food, to cite only a few examples, are made with an astonishing precision but substantial social consequences. The social security net may need to be enhanced for this purpose.

Never before has our reliance on automation been so strong. The increasing knowledge of robotics and artificial intelligence will invade all walks of life. A term nowadays often used and quite debated to describe this development is that we are approaching a "singularity" in the next two decades. The singularity concept portends that the impact of these accelerated changes will be so profound that they will provoke a rupture in the fabric of our society. It appears that the term "singularity" was earlier used during a conversation in the late 1950s between Stanislaw Ulam and John von Newman, in reference to the "ever accelerating progress in technology and changes in the mode of human life, which gives the appearance of approaching some essential singularity in the history of the race beyond which human affairs, as we know them, could not continue."

The concept of "singularity" has since been widely used in scholarly papers and was elaborated on by Ray Kurzweil in one of his bestselling books, *The Singularity Is Near*. In what he describes as the Six Epochs of Evolution, the singularity will begin in Epoch 5, the merger of technology and human intelligence, which will continue into Epoch 6 and expand thereafter. The timeframe for the occurrence of such an event is 2045. It is my assessment that Epoch 5 is underway.

As technology is unstoppable, opinions vary greatly among scholars about the implications at the individual and collective levels. A closer analysis of this issue may reveal that opinion differs in segments of the general public according to religious beliefs and resistance to change, in spite of the abundant information available. Each of these three factors will play an important role in the resentment, indifference, or positive expectations that arise in response to the forthcoming development; the religious beliefs factor will be dealt with separately in Chapter VII as part of a broader issue of where humanity is heading. This is not to rule out vigilance or a need to be informed about these issues. These concerns lead to deep thinking about the greatest challenges facing humanity. As new technologies diffuse, penetrate, and permeate our society, people will react by adapting and deriving benefits from the new opportunities or by rejecting them. One of the concerns often raised is that those opportunities will only be available to those who can afford them.

The Gap between the Haves and Have-nots

An economically seamless society has always been sought, explicitly or not, throughout the many revolutions and political reforms in our history. Alan Maass, editor at socialsworker.org, in the article "What Really Happened in 10,000 B.C.?" (issue 667, March 28, 2008), traced the beginning of the haves and have-nots since the emergence of the sedentary society in which the collectively agreed upon rule of distribution of tasks and compensation prevailed. He expressed the view that the elites (the haves) involuntarily acquired privileged positions through the passing of time. In other words, the rise of a class society was an unintended consequence of developments in the ways humans produced the necessities of life. Karl Marx believed that the interest of the (production) owner, the elite, or the bourgeoisie in maximizing profit is in opposition to the interest of the workers, who aim to maximize their wages and benefits. In Karl Marx's view, this clash of interests creates a contradiction in the capitalist system, with both parties apparently being unaware of this antagonism. To this end, I will only offer this quote from Adam Smith, repeated many times in this book: "The reciprocal relationships that people voluntarily establish, channel self-interest to mutual advantage and promote a prosperous social order."

Disposable income varies greatly between individuals and, in countries where democracy prevails, taxation, education, pension and health reforms can be seen as instruments to achieve a fair balance in the purchasing power and fulfilment of individual needs. Whether this goal can be totally achieved remains a moot point because it is a moving target; in addition, income distribution means different things to different people and groups of people with different political ideologies. It is not the aim of this book to expand on this fluid subject, but in the context of technological development outcomes manifested in the availability of new products and services, the difference between the haves and have-nots has a different meaning. It covers many grounds: well-being, wants versus needs and, more importantly, affordability. It is in this latter category that the challenges arise.

Fifteen years ago, I purchased a desktop computer, which today I consider rudimentary, for $5,000. Today, I can buy a computer with much higher performance for 10% less. The same applies to pocket calculators and portable radios, which can be bought for a fraction of their original price. Indeed, technology has combined calculators, music, and much more in the mobile phone. Since 2015, it is not uncommon to find more than one desktop computer, laptop, and mobile phone in a household, which was out of the question decades ago. Today most residences are equipped with cooking units that include microwaves, and with refrigerators, televisions, and laundry appliances. In the case of television, the "black-and-white TV" has shared the fate of the typewriter: gone forever. Color TV is taken for granted. The affordability of these items seems to be a less divisive issue.

My point is that initially, all innovations tend to be expensive, not affordable for everyone. When first put on the market, they are affordable for only a few, before the interplay of competition, demand, and supply forces. The same goes for certain medical interventions for cosmetic reasons, such as facelifts, hair replacement, etc. It has also been the case with accessibility to the internet. (This latter point will be addressed in more detail in Chapter VI.) Then prices go down, accompanied by substantial improvements to the products or services.

We all have been spoiled by the goodies of technology to a point where some products and services, such as indoor extra-clean air and internet accessibility, are about to join the list of basic rights rather than being privileges accessible only to a few. After all, this is the purpose of progress: raising our standard of living.

It also appears that the root of social frustration more often than not is consumers' confusion about their wants and needs, and the growing instant-gratification phenomenon. As the inhabitant of a developing country, for instance, you may want a 5GB cellular phone but do not need one if the infrastructure does not support its utilization to the fullest. In most developed countries, we want everything now. This statement is not intended to be a value judgement of a cultural phenomenon; after all, this is part of living in a free-market system. In developing countries, globalization is pursuing its course of bringing products and services to consumers but at a slower pace. The reasons for this delay are multiple: weakness in or absence of the appropriate infrastructure, lack of distribution channels, excessive custom duties, lower income levels, and so on. But as a wise man said, there is no problem without a solution. In this case, time seems to be the best remedy. Overall, I believe that it is reasonable to say we are all better off than our parents and many steps ahead of our grandparents.

In the meantime, there might be a way to facilitate the accessibility of certain products and services so that they are available to many. Technological innovations can be infused and diffused in different forms. Wikipedia defines frugal innovation as the process of reducing the complexity and the production cost of products to meet a market requirement. Such products and services need not be of inferior quality but must be provided at an affordable price. Frugal innovation, translated as *jugaad* in Hindi, is currently taught in many universities, including those in the United States, and is an approach adopted by large multinational corporations aiming at doing more with less. In the book *Jugaad Innovation: Think Frugal, Be Flexible, Generate Breakthrough Growth*, by Navi Radju Radjou, Jaideep Prabhu, and Simone Ahuja, there are many stories of fascinating achievements in Brazil, China, India, and Costa Rica illustrating the financial and social rewards of applying a systematically frugal approach to product delivery. From mobile banking to mobile hospitals, it is evident that creativity in the adaptation of innovations to local markets is not only possible but also a good business practice.

Whether countries are able to develop the growth and employment potential of imported technologies depends on the creativity and the absorptive capacity of domestic firms. This approach could well be a template for developing countries to modify foreign innovations to better serve their markets.

What Are the Recurring Themes in the History of Technology?

Leitmotif is the Anglicization of a German word meaning repetition, particularly in music, of a particular (changing) theme to unify the various components of a piece. Coined by the famous German composer Richard Wagner in the 19th century, this term has been used in other fields to illustrate the recurrence of any particular fact. In technology, throughout the course of time, there appear to be four major areas of recurring activities: agriculture, health, education, and energy seem to be our constant preoccupations. Although our current activities are spread in these four areas, the leitmotif of our species has also been the conquest of outer space, as part of our innate curiosity. But first, let's examine how the four components act as a refrain in the never-ending technological development song.

Agriculture

From the lifestyle of wandering hunter-gatherers to sedentary agrarian-shepherds, almost 200,000 years elapsed before the first human agricultural settlement. One can imagine the hardship that they faced, starting with the capture and domestication of animals; their daily food supply; the prevailing environmental risks; the fragility of animal shelters against predators; the learning process regarding inedible grains, coupled with an inadequate supply of edible food. Using the philosopher and historian Thomas Hobbes's words from another context, life was nasty, brutish, and short. The Neolithic period, which is associated with the first agricultural revolution, occurred in 8000 BC and saw the beginning of agricultural communities. The most recent information available seems to confirm the rudimentary tools, by today's standard, used at that time for the farming of crops. Sickles were invented in 2000 BC, while the hoe and plow came much later, subsequent to the taming of horses and the use of oxen. Improvements in tools and implements increased the yield in crop farming and in turn forced the relocation of farms when the nutrients of the soil in one location were exhausted. It was mostly monoculture farming, comprising very few types of grains, such as wheat and barley. The challenge was the need for green pastures for animals for human consumption. In the beginning, agriculture and agricultural implements were amazingly simple. The crop rotation method to rebalance the soil's nutrient composition was introduced in 6000 BC, and irrigation was introduced in the following millennium.

But the technology of the Neolithic period became more evident as the expansion of communities took place. In this new settlement or lifestyle, food was more available, and much more time could be spent on other activities to improve life, such as pottery and weaving, and polishing stones for household use. This was perhaps the environment that was conducive to the early blooming of technological development, but nothing substantial really happened until the Bronze Age. The hoe and the plow were among the first agricultural implements of the first agricultural revolution that occurred in 5000 to 6000 BC along the banks of the Tigris and Euphrates Rivers, now Iraq. An expansion of the sedentary settlement required more domestication of plants and animals and more human effort. The addition of one of the domesticated animals removed some of the burden from humans in the agricultural field operations. Harnessing the pulling power of horses was indispensable, considering the large areas to be plowed, seeded, and harvested. At that time, necessity was really the mother of invention. The Middle Ages brought the invention of horseshoes and nails, and horse collars. Larger crop production occurred because of the development of more villages, but there wasn't much crop variety until the third agricultural revolution and the mechanization of agriculture.

Today, agriculture is widely studied and crop management has become a global science. Biotechnology is becoming the hope of humanity to counteract the impact of climate change upon agriculture. The increased quantity of carbon dioxide (CO_2) in the atmosphere in the last decades has affected the quantity (yield) and the nutritive quality of crops in terms of iron and zinc concentration, potentially exposing the global population to famine or malnourishment in certain countries. The increased CO_2 level affects the most commonly known crops, such as wheat, barley, oats, rice, and cotton, basically 85% of all crops (10). In the words of the journalist Alan Weisman, author of *The World Without Us* and many other books, "we must always be aware that any species that overstretches its resource base suffers a population crash. Limiting our reproduction would be hard, but limiting our consumptive instinct may even be harder." Scientific findings reported in *Nature* suggest "that breeding for decreased sensitivity to atmospheric concentration could partly address these new challenges to global health."

There is also another challenge to be addressed in terms of limitation in the food supply. According to the OECD report entitled *The 21st-Century Technologies: Promises and Perils of a Dynamic Future*, of the 30,000 edible plants grown, only six supply 90% of human nutrition. Another report by Live Science states that just three crops (wheat, rice, and maize) account for more than 60% of the calories consumed by approximately seven billion people across the world today. Factors preventing the consumption of the rest of the edible plants are bad taste, amount of energy required to make them edible, unattractive appearance, and social taboos. It is therefore reasonable to expect biotechnology to offer greater diversity in the food supply by making some of these plants usable by society. The global food supply system appears to be very fragile; this may be one of the factors explaining the growing practice of monoculture.

Recently, genetically modified food (GMF) has garnered a lot of attention and a mixed reception but also has become a debatable issue around the globe, particularly in Europe and America. Many concerns are raised at the ethical and ecological levels. Among them are: trans-genetic operations, meaning the mixing of genes from different species; the consequences of GMF for health; disequilibrium in ecosystems; and "unnaturalness." Many people are also concerned about extreme allergies or intolerances to trace amounts of other substance(s) in the composition of genetically modified products.

The practice of spraying fruits and vegetables for conservation purposes has also attracted a lot of attention, and this commercial practice has its pros and cons. Labelling is a further issue which needs to be considered. Consumers must know the components of the products they are buying, although one can question the weight of this factor in the purchasing process, considering the fast pace of our buying environment. I tend to concur that for the sake of transparency, the full composition of the product being purchased is needed, particularly where health issues are concerned. Nevertheless, the snowball effect caused by people reacting to simple hearsay is troubling. Intuition seems to be a driving factor in most people's rejection of GMF and, on the basis of scientific evidence accumulated thus far, it would be logical to expect people to balance this intuition with a more analytical approach. One of the cornerstones of GMF is food security, and over time this will become a major issue worldwide. Figure II.5 displays the results of an in-depth study carried out in India on the impact that genetically modified cotton has had on cotton farmers. *Bacillus thuringiensis* (Bt) is a soil bacterium that produces insecticidal toxins. I invite the reader to at least browse this article, particularly the survey methodology, which adds to the credibility of the study (11).

Four issues are discussed in this article: food production, safety, and quality; and socioeconomic impact on farmers. The study reports more yield per area, less use of pesticides, more economic benefits, and increased levels of nutrients in the food consumed by farmers who adopt Bt cotton. Fig II.6 shows the market penetration in the adoption of Bt cotton, from a slow beginning in 2002 to a greater acceptance in 2008. Genes from Bt can be inserted into crop plants to make them capable of producing an insecticidal toxin, therefore making them resistant to certain pests.

Farm households	2002	2004	2006	2008	Total
Adopters of Bt	131	246	333	375	1085
Non-adopters of Bt	210	117	14	5	346
Total	341	363	347	380	1431

doi:10.1371/journal.pone.0064879.t001

Figure II.5 Penetration of Bt cotton in farming in India

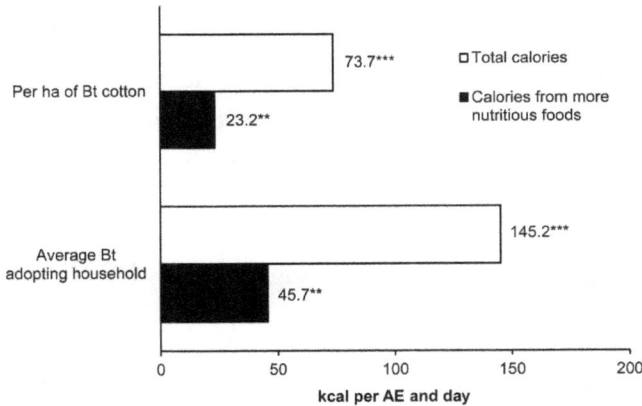

Figure II.6 Calorie consumption by adopters of Bt cotton, AE (adult equivalent)

Figure II.6 shows an increased level of calorie consumption by those adopting Bt cotton.

As the yield per hectare increases with Bt cotton, so does the revenue of the farmers. The increase in disposable income allows these farmers to diversify their diet and consume more nutritious food in addition to the traditional cereals, which are rich in carbohydrates but less so in protein and micronutrients. Regardless of this success, there is still a debate about the use of fewer pesticides in the cultivation of Bt cotton.

Health

Life expectancy has risen in all countries around the globe. One can see an ascending trend in human health improvements, starting from the Enlightenment period and increasing to the present day. Consider the following facts. As shown in Figure II.7, the global weighted average life expectancy at birth was around 30 at the beginning of 1750, compared to 60 in 1900 and 80 at the beginning of the current millennium. It is reasonable to suggest this life extension improvement is the combined result of extended vaccination, sanitation, and the use of antibiotics against infectious diseases. The x-axis shows the cumulative share of the world's population. All the countries of the world are ordered along the x-axis, ascending by the life expectancy of the population. The y-axis shows the life expectancy of each country. In 1800 (represented by the red line), the countries on the left—India and also South Korea—showed a life expectancy of around 25. On the very right you will note that in 1800, no country had a life expectancy above 40 (Belgium had the highest life expectancy at just 40 years). In 1950, the life expectancy of all countries was higher than in 1800 and the richer countries in Europe and North America had life expectancies above 60 years. Over the course of modernization and industrialization, populations' health improved dramatically, but half of the world's population, including in India and China, made little progress.

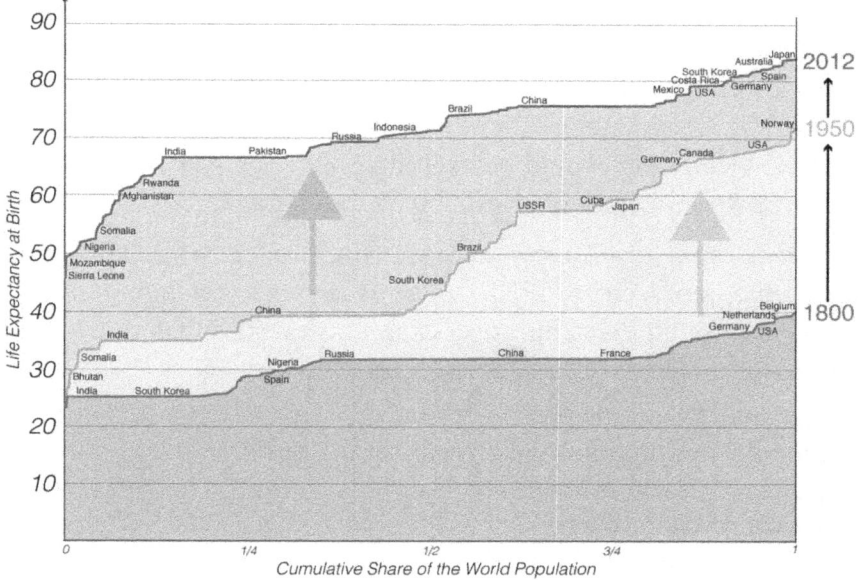

Figure II.7 Life expectancy of the world population
http://ourworldindata.org/data/population-growth-vital-statistics/life expectancy/
Max Roser (2015) – "Life Expectancy." Published online at OurWorldInData.org.

Therefore, by 1950 the world population was experiencing highly unequal living standards clearly divided between developed countries and developing countries. Between 1950 and 2012, that division ended. The developing countries that were worst off in 1950 have achieved the fastest progress, while some countries (mostly in Africa) are falling behind. Many of the former developing countries have caught up, resulting in a dramatic reduction in global health inequality. The world developed from equally poor health in 1800 to great inequality in 1950, then back to more equality today, but equality on a much higher level.

In this context, considering the current state of medical technology, can longevity (living to be over 100) be achieved? The physicist and writer Robert Freitas Jr. believes that life expectancy could be extended beyond 150 years by eliminating 50% of a list of preventable medical conditions. In a very comprehensive lecture entitled "Extreme Life Extension," delivered on November 16, 2002 in Newport Beach, CA, he stated that longevity could be achieved when life extension reached a double-digit increase every decade: "If we could get to a rate of 10 years of life extension per decade, then medical technology would be extending life exactly as fast as we're aging, postponing natural death, on average, indefinitely."

The gerontologist Aubrey de Grey in his "Strategies for Engineered Negligible Senescence (SENS)" (15) maintains that aging can be reversed through the application of seven strategies that address the cellular or molecular damages accumulating with age:

- Cell loss, tissue atrophy
- Nuclear [epi]mutations (only cancer matters)
- Mutant mitochondria
- Death-resistant cells
- Tissue stiffening
- Extracellular aggregates
- Intracellular aggregates

Although successful in animals, no application has yet been made on humans. Many serious studies to understand the causes of aging are ongoing in most developed countries, such as those of the National Institute of Health and the longitudinal studies on aging in Baltimore (BLSA) in the USA, and in Canada, the UK, Australia, and Japan.

Outside of ethical or metaphysical considerations, which will be covered in the next chapter, could longevity be considered desirable? Yes. I suggest that non-receptivity of a prolonged, healthy life beyond the socially accepted 100-year set limit is mostly associated with the Tithonus syndrome. In Greek mythology, Tithonus was a mortal who was granted eternal life by Zeus but forgot to ask for eternal youth. As a result, he became increasingly debilitated and demented as he aged. Fortunately, this is not a scenario envisaged by most gerontologists. Longer life does not mean longer dying. The real trend is towards a longer and more active life. It seems that preconceived ideas and myths nudge people to a particular position, precluding them from making any in-depth analysis of the issue at stake.

The reader interested in knowing more about the myths that people have about aging should spend some time on the following site of the well-known gerontologist João Pedro de Magalhães, PhD (17). On this site, you will find words and expressions containing what I would call debate boxes. They raise high-interest issues designed to stimulate discussions to correct deep-grounded fallacies about aging. For example, you will get the answers to some of the most frequent questions or issues about aging:

- Myth #1: Aging is natural and we should not fight it.
- Myth #2: What is the point of extending life if we are old?
- Myth #3: A finite lifespan is best enjoyed.
- Myth #4: Why should life be better than death?
- Myth #5: Not everyone would benefit from a cure for aging.
- Myth #6: Economic disaster would result, with the collapse of healthcare.
- Myth #7: Overpopulation would lead to a total catastrophe.
- Myth #8: Human trials of an "anti-aging pill" would be dangerous.
- Myth #9: Humankind as we know it would change.
- Myth #10: We should have other priorities on Earth.
- Myth #11: Overall, trying to cure aging is ethically wrong.

I am hoping that I have stimulated your curiosity enough to visit the site mentioned above.

Curative medicine continues to make substantial inroads in parallel with the development of regenerative medicine. Considering the current state of medical technology, it does make sense to give priority to research projects on longevity, because the real causes of aging are still unknown. In the words of Sebastian Sethe and João Pedro de Magalhães in *Ethical Perspectives in Biogerontology*: "The underlying cellular mechanism of aging and the process(es) driving the aging process are still poorly understood. So far, the two prevailing theories are: (a) that aging results from pre-determined mechanisms—program-based theories—usually with an element of genetic regulation and (b) that aging results from random or stochastic damages—damaged-based theories—including the free-radical theory of aging, which suggests that a gradual buildup of oxidative damage with age drives the aging process, and the idea that DNA damage accumulation with age causes the physiological and functional decline we call aging."

From the foregoing it would appear that an approach about solving old-age-related diseases is a reasonable approach to address aging. However, despite the causation between certain diseases and the deterioration of a related organ, such an approach would be only partial for failing to address the causes of aging. This whole endeavor would confirm a cynical joke: "The operation was a success, but the patient died." Speaking against the current piecemeal approach to aging, the eminent sociologist S. Jay Olshansky, in "A Wrinkle in Time: A Modest Proposal to Slow Aging and Extend Healthy Life," wrote: "While we can extend life in aging bodies through behavioral improvements and medical treatments, the time has arrived to acknowledge that our current model of reactive medicine, of trying to treat each separate disease of old age as it occurs, is reaching a point of diminishing returns" (19).

Can substantial life extension be achieved and, if so, how long will it take? It is just a matter of priority and funding. In this respect, the goal of Calico must be underlined. Currently part of the new Alphabet conglomerate along with Google, Calico is one of the very few private-sector companies to have recently set a goal to cure aging and ultimately extend life. Having said that, it is also fair to extrapolate a few implications of life extension beyond 100 from a financial, judicial and social standpoint. In financial services, the underwriters of life insurance, accident, and sickness coverage would have to offer a panoply of insurance types to meet the growing market needs of centenarians. This is not raising a red flag. There is nothing in the above speculation that the actuarial science cannot address. Equally, the investment horizon will need to be modified to reflect a much later maturity date. This modification may not affect pension plans, as the contributions will be extended for a longer period of time. In the judicial system, life sentence in prison will have another connotation, surely more costly for governments, as inmates will need to be incarcerated for a longer period of time. In social services, the retirement benefits will have to be paid for a longer period of time. At this time, I am doubtful that people will choose to stay on the labor market much longer, because of the panoply of activities not restricted by age, including sports and a leisure society lifestyle. Conceivably, permanent education will become a fact of life, as people by choice may change career many times or continue to study over the course of their long life.

Energy

The development of our economic system and the continuation of our individual well-being rest on a sustainable production of energy. Since the middle of the preceding century, the use of fossil fuel, one of the engines of the industrial revolutions, has been skyrocketing. A

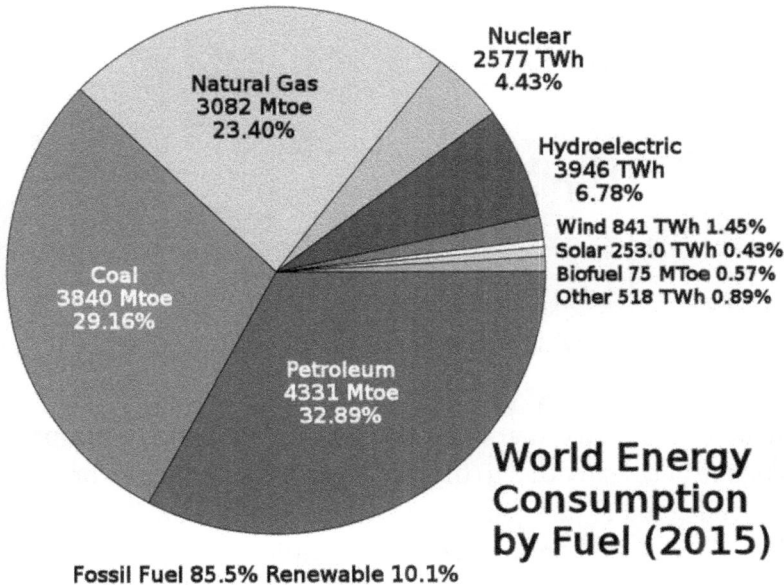

Figure II.8 World energy consumption by fuel Source: Wikipedia.org https://commons.wikimedia.org/wiki/File:World_energy_consumption_by_fuel.svg

short reduction occurred between 1971and 1973 during the petroleum crisis, but that is not visible in the above graph. This in turn has raised the interest in renewable energy. As shown in Figure II.8, the combined consumption of petroleum, coal, and natural gas amounts to 11,253 million tons of oil equivalent (Mtoe) or approximately 85.5% of the total energy consumed (for the nuclear component: 2577 TWh = 0.2215821 Mtoe). The use of fossil fuel in land, maritime, and air transportation and in the manufacturing industry will prevent its elimination overnight. Although the advent of electric cars may expand to other land transportation areas, this innovation will not make a significant contribution to carbon dioxide reduction in the atmosphere unless more stringent regulations are put in place.

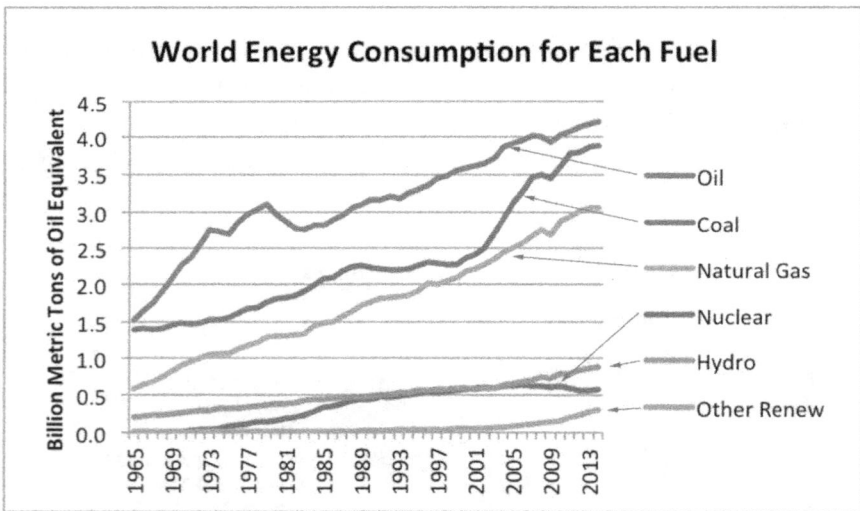

World Energy Consumption for Each Fuel

Figure II.9 World energy consumption

Source: ourfiniteworld.com – World per capita energy consumption by fuel, based on BP 2016

Coal is the fossil fuel that drove the first industrial revolution. Its use has since been extended to the electricity generation, the production of steel, cement, and paper, and other uses in many countries, including China, the United States, and India. These are products that we enjoy or will find it difficult to get rid of without severely impacting the economy.

Natural gas, the third category of fossil fuel, has proven to be the least polluting. Trillions of cubic feet of gas trapped between shale rocks may be recoverable. However, some environmental concerns have been expressed about hydraulic fracking, the method used to release the gas by high-pressure injection of water, sand, and chemicals.

With the commissioning of the Obninsky Nuclear Power Plant in 1954 (in the then USSR) and Shippingport Atomic Power in 1957 (USA), nuclear power appeared to be a new promising source of electricity generation. Basically, a nuclear power plant functions like a thermal power station. The heat from nuclear fission causes a coolant to produce steam, which activates a steam turbine to power an electric generator. Electricity is then produced without harmful effects to the environment. However, this not without risks—two of which are potential radioactive contamination in the extraction and transportation of uranium, the main component to fuel the process, and the complexity of the constant monitoring of such a system, which increases the probability of human error. Accidents such as those at Three Mile Island (USA, 1979), Chernobyl (Ukraine, 1986), Pickering (Canada, 1994), Manche (France, 2002), Sellafield (UK, 2005), and Fukushima (Japan, 2006) have tainted the reputation of nuclear power plants as safe and reliable sources of electricity, in spite of the corrective measures taken.

The use of fossil fuel is not without negative consequences. As mentioned earlier, fossil fuel cannot be eliminated in the foreseeable future. I therefore suggest that research and development spending be increased to counteract the negative effects of this source of energy until it can be replaced or substantially reduced by renewable energy sources. In the meantime, based on the recent Paris Accord of the United Nations Framework Convention on Climate Change (UNFCCC), there is reason to expect a "stable" environment if the commitment to keep global warming well below 2.0 °C by the end of this century is met by cutting greenhouse gas emissions. Fortunately, the growing concern about carbon emissions and the resulting political pressure will lead us in the right direction.

As shown in Figure II.9, renewable energy, the environmentally friendly source of energy, is progressing at a turtle's pace. The storage, production, usage limitations, and appropriate infrastructure requirements are just a few of the impediments preventing the widespread of renewable sources. Gail Tverberg (14B) in a documented article has estimated that it will take 860 years to completely eliminate the use of fossil fuel. This reminds me of the Russian astrophysicist Nicolai Kardashev's scale of civilization (14), according to which developed countries are at level 0. It means that we will continue to draw our energy from underground, from dead plants and animals, for a long time.

However, many countries are actively working towards the goal of eliminating or reducing their use of fossil fuel. Denmark has set a goal of meeting 100% of its energy needs from renewable sources by 2050.; they are currently at 43%. Germany has set a goal of meeting 45% of its energy needs from renewable sources, making it a leader in solar energy and in the conversion of municipal solid waste (MSW) to energy. Renewable energy will provide 14% of its gross electricity consumption by the end of 2015. New Zealand meets 72% of its energy needs from renewable sources. Nevertheless, it is not always appropriate to copy the technological advancement of a particular country, because the geographical conditions and endowments vary from one place to another. For instance, Canada is well positioned to capitalize on hydro power, unlike some countries in the Middle East.

In at least 19 countries around the world, the production cost of solar energy has already reached grid parity with the traditional electricity production cost. Grid parity occurs when an alternative energy source can generate power at a levelized cost of electricity that is less than or equal to the price of purchasing power from the electricity grid. According to Bloomberg Business, solar energy has already reached grid parity in 10 states that are responsible for 90%

of solar electricity production in the USA. The Deutsche Bank projected that rooftop solar PVs (photovoltaic cells) would reach grid parity in the 50 American states by 2016.

Waste to energy (WTE) as another source of renewable energy is quite promising. Do Not Waste the Waste is an attractive motto to put people in the proper mindset. Indeed, the useless is becoming increasingly useful. There are currently 87 WTE conversion facilities located in 29 American states. In a publication of the World Bank ("What a Waste," March 2012, Number 15), it is forecast that the volume of municipal waste will increase from 1.3 billion tons per year in 2012 to 2.2 billion in 2025. In most European countries, landfill, when available, has very restrictive requirements. The strategy of converting municipal solid waste to energy is a win-win alternative, particularly for developing countries, because the technology exists, keeping in mind that recycling and WTE are not mutually exclusive.

Urbanization is a demographic trend common to all countries around the globe. I submit that undeveloped and developing countries could also benefit from the application of WTE technology because of its positive impact on health, education, and tourism. In the Caribbean, Dominican Republic (DR), Jamaica, and many others have set an energy roadmap to reduce their dependence on fossil fuel. Currently, many public buildings in DR are equipped with solar panels, saving $50 million annually. In Central America, Costa Rica aims at becoming carbon neutral in the next decade. In South America, Brazil gets at least 75% of its energy supply from renewable sources. Argentina has passed legislation to encourage industrial companies to obtain 20% of their electricity needs from renewable energy sources.

These few examples are evidence that there is a global trend towards reducing carbon emissions in the environment by using wind turbines, solar panels, biomass, or any combination thereof. A combination of solar panels and wind turbines can be used to provide electricity to small communities, as described at the end of this chapter (14A).

It is accepted that energy is one of the most crucial factors for economic development. Since the sustainable production of clean energy reflects the technological advancement of a society, the use of the above-described renewable energy technologies will contribute to attracting investments as part of the globalization process and subsequent economic developments. It is my hope that renewable energy growth will continue and at a rate exceeding energy demand so we can avoid reverting to fossil fuel in the future.

Education

In 1814, James Pillans, headmaster of the Old High School of Edinburgh, Scotland, invented the blackboard. Quite a celebration ensued over this new pedagogic tool that could transmit knowledge. The blackboard was then perceived as a giant step towards the democratization of knowledge. In 1837, the Massachusetts Board of Education dedicated a substantial portion of its annual report to praising this invention. Pillans has been remembered as a reformer of education. Needless to say, this information technology changed the teaching and learning process. Today the user-friendly character of most educational software, coupled with the improved price-to-performance ratio of laptops, tablets, and mobile phones, play the same role. They accelerate the delivery of education beyond geographical limitations. Information technology facilitates academic research and reduces crowding at libraries. Because of this tool, universities have become more enablers than providers of knowledge, nurturers of creativity, loyal advisors/partners for the private sector. But there is a broader role that information technology is playing and will continue to play in the foreseeable future: the interuniversity collaborative spirit in online education delivery by the massive open online course (MOOC) initiative around the globe. This vehicle democratizes education and makes it possible for an individual to increase his/her knowledge through personal initiative. Free up to now, people can select from a broad variety of courses from a number of well-recognized colleges and universities. In addition to being a great windfall of the digital age, the evolution of distance learning has not reduced university campus populations. At the macro level, it seems that lifelong learning is gaining momentum; society is recognizing that the population needs to adapt to new market conditions and/or people's motivation to learn.

The preceding section addressed the available options for the general public to acquire or increase competence in various disciplines. However, the ramifications of education are far broader. It prepares the younger generations for the labor force and participation in democratic society. As such, education is one of the vehicles that can reduce inequality and increase voter turnout by by raising a better understanding of the political parties' platforms. I have chosen not to address the issue of civic education. It is one of those issues widely recognized as important but potentially controversial in its implementation (20).

Many actions have been taken to facilitate access to education and the marketability of colleges and universities to students. For instance, the student loan program created in the preceding century (1939 in Canada, 1958 in the United States) has been subject to many improvements over time. However, it has recently become a major burden, equivalent for some to the load of a 20- or 30-year mortgage, if not more. Perhaps repayment of the interest only could be an alternative to consider, the capital being the country's investment in the regeneration of its labor force.

The acquisition of skills from a practical viewpoint is often one of the causes of graduate students' unemployment. Most employers, quite rightly, expect hands-on experience, particularly for high-skills work. Indeed, there is again a mechanism in place to address this deficiency. Most Canadian and American colleges and universities have cooperative or internship programs, allowing students to practice their newly acquired skills in a real paid work environment. High academic performance and efficiency on the job are two different things. In addition to acquiring experience, and earning an income to pay for some ordinary student expenses, such programs may also facilitate employment. I suggest that measures be taken to encourage employers' participation in co-op and internship programs, such as a tax credit for the employer and information from the Bureau of Statistics to colleges and universities on the labor market's needs.

Suggested Readings

(1) "Big History," https://en.wikipedia.org/wiki/Big_History

(2) Fred Spier, *The Structure of Big History: From the Big Bang until Today* (Amsterdam: Amsterdam University Press, 1996)

(3) David Christian, *Maps of Time: An Introduction to Big History* (Berkeley: University of California Press, 2004)

(4) David Christian, *The History of Our World in 18 Minutes* http//www.ted.com/talks/david_christian_big_history?language=en

(5) George Woodcock, *The Tyranny of the Clock* https://www.acsu.buffalo.edu/~rrojas/TyrannyofClock.html

(6) Jeremy Dittmar, *The Impact of the Printing Press* http://voxeu.org/article/information-technology-and-economic-change-impact-printing-press

(7) Howard Markel, "The Real Story behind Penicillin" http://www.pbs.org/newshour/rundown/the-real-story-behind-the-worlds-first-antibiotic/

(8A) "Charles Goodyear: Inventor of Vulcanized Rubber" http://www.american-inventor.com/charles-goodyear.aspx

(8B) "Alexander Fleming's Discovery of Penicillin" https://www.acs.org/content/acs/en/education/whatischemistry/landmarks/flemingpenicillin.html#alexander-fleming-penicillin

(8C) "Circulation: Cardio-Vascular Intervention" http://circinterventions.ahajournals.org/content/4/2/206

(8D) "Multiple Discovery" https://en.wikipedia.org/wiki/Multiple_discovery

(9) "How Does the Electoral College Work?" http://people.howstuffworks.com/question472.htm

"Jugaad Innovation: Think Frugal, Be Flexible, Generate Breakthrough Growth," *Stanford Social Innovation Review* http://ssir.org/articles/entry/jugaad_innovation_think_frugal_be_flexible_generate_breakthrough_growth/

(10) "Increased CO_2 Threatens Human Nutrition" http://fcrn.org.uk/research-library/increasing-co2-threatens-human-nutrition

(11) Matin Qaim and Shahzad Kouser, "Genetically Modified Crop and Food Security" http://journals.plos.org/plosone/article?id=10.1371/journal.pone.0064879

(12) "Cosmic Time Scale" https://en.wikipedia.org/wiki/Cosmic_Calendar

(13) GMO Compass, "Cotton Is More than Just a Fiber for Textiles…" http://www.gmo-compass.org/eng/grocery_shopping/crops/161.genetically_modified_cotton.html

(14) The Nicholai Kardashev Scale of Civilization futurism.com/the-kardashev-scale-type-i-ii-iii-iv-v-civilization/
(A) A Residential Example of Hybrid Solar–Wind energy http://facultyfiles.frostburg.edu/engn/soysal/Activities/Publications/IEEE_PES08GM(2007-11-24).pdf
(B) The "Wind and Solar Will Save Us" Delusion ourfiniteworld.com/2017/01/30/the—wind-and-solar-will-save-delusion/

(15) Aubrey de Grey, "Strategies for Engineered Negligible Senescence" http://www.senescence.info/sens.html

(16) Sebastian Sethe and João Pedro de Magalhães, "Ethical Perspectives in Biogerontology" http://pcwww.liv.ac.uk/~aging/ethics13_aging_life_extension_research.pdf

(17) Education and information about the science of aging www.senescence.info

(18) National Human Genome Research Institute, "Germline Gene Transfer" https://www.genome.gov/10004764/germline-gene-transfer/

(19) Kurzweil R. *The Singularity is Near*. New York: Penguin Books, 2006.

(19) S. Jay Olshanksy, "A Wrinkle in Time: A Modest Proposal to Cure Aging and Extend Health Life," *Slate* http://www.slate.com/articles/technology/future_tense/2010/11/a_wrinkle_in_time.3.html

(20) "Civic Education" https://plato.stanford.edu/entries/civic-education/

CHAPTER III

Socioeconomic Impact of Technology

*"Technologies are not simply inventions which people
employ but are the means by which people are invented."*
—Marshall McLuhan

The preceding chapters covered the progress and nature of technology, but its ramifications in our current society form a fascinating story that remains to be told—the transformation of our
society. As per the timeframe set in this book, the review started
at the Renaissance, but three inventions before this period (3) have
had important ramifications even to the present day. We would be
living in a different world without the momentum created by these
innovations: gun powder (1249), the spinning wheel (1290), and the
mechanical clock (1300).

Gun Powder
The history of gun powder goes as far back as the Tang dynasty in
China (AD 618 to 907) and the search by ancient alchemists for an
elixir of life that would render the user immortal. The miraculous
potion was never found. However, one important ingredient common to the many failed potions—saltpeter, also known as potassium

nitrate—was found to have explosive capacity. There is no need to comment on the fate of the alchemists' first trial. What an irony! Instead of providing immortality, this magic potion could end life. Although potassium nitrate was invented during the Tang dynasty, the Song dynasty is credited for making known the formula of explosive gun powder. It consisted of 75 parts potassium nitrate, 15 parts charcoal and 10 parts sulfur. Gun powder was an important if not determining factor in the Chinese end to the constant invasions of the Mongols. Concerned about the potential proliferation of its technology, the Song dynasty in 1076 prohibited the sale of potassium nitrate to foreigners. However, this prohibition could not be enforced because of commercial exchange with India and Europe, and by the 13th century, this technology was found in rifles, handguns, and cannons. Gun powder has completely changed the way war is fought. It has moved the fight from castles to the open field, leading to the formation of infantry and artillery.

It is also worth mentioning at least two civil applications of gun powder in China—namely, the excavation of canals for irrigation works, and fireworks. Recent modifications of the chemical formula of gun powder have allowed for spectacular displays of intense colors and forms at nights of national celebration.

The Spinning Wheel
The spinning wheel, which later evolved into the spinning jenny, radically transformed the European textile industry from:

- a home-run business to a factory enterprise,
- a family business to a diverse-entity business, and
- scattered suburb businesses to a central location (i.e., cities)

This socioeconomic transformation cannot be attributed solely to the spinning jenny. Textile merchants since the Renaissance had been under pressure to deliver their commodities to clients, a pressure that was thereafter compounded by the efficiency in the agricultural production of cotton, the raw material for textiles during the First Industrial Revolution. The merchants were at the mercy of the (family) workers, who worked at their own pace in order to meet their own immediate family needs. The decision to move textile fabrication to a central location can then be explained not only from a logistical viewpoint—less traveling time for the merchants to the workers' home—but also from a business efficiency viewpoint, because production would be higher in a controlled environment. This latter goal was achieved in part via a regimented work schedule from morning to evening, not necessarily at the will of the workers, as well as simplified work processes in which the necessary equipment was provided by the business owners.

Thus, the factory system was born and, soon thereafter, affluent cities and residences, as workers and business owners wanted to be close to their work. There is one important fact to note in the business transfer from home to factory: machines reduced the need for high-skill workers by bringing in low-skill workers for "mass" production. The reverse was seen thereafter until the present day.

The Mechanical Clock

Another machine that has had a tremendous influence on present-day society is the mechanical clock. Before this invention, time and human activity were loosely measured according to the pattern of sunrise and sunset. With the invention of the clock, time "took on the character of an enclosed space: it could be divided, it could be filled up, and it could be even expanded by the invention of labor-saving instruments," as *The Lewis Mumford Reader* describes. One can see the stepping-up of efficiency, particularly in this more controlled work environment, which signaled the "calibration" of human capability with the ever-perfected machine output. This would be amplified later by F.W. Taylor as part of the First Industrial Revolution. Needless to say, there has been some abuse of this efficiency in terms of longer hours of work. Time has become not only a commodity that can be bought and sold, but also the regulator of humans and other machines. Even if efficiency is a desirable trait in society, one can see the resulting pain caused by the quest for more.

The Canadian writer George Woodcock in *The Tyranny of the Clock* recognizes the need for machines in "reducing unnecessary labor to a minimum" but also states that the domination of man by a creation of man is even more ridiculous than the domination of man by man.

The Movable Type Printing Press

We now return to the mid-15th century, by which time almost one-third of the population of England could read and write. In France, this century saw the beginning of the influence of François Rabelais, who is considered one of the creators of modern European writing. A rebirth of learning had found traction with the invention of the movable type printing press by Johannes Gutenberg in 1450, the first vehicle for democratizing knowledge. From a humble production of 3,000 pages per work day, the movable type printing press would later produce 150 to 200 million pages per day, as it became more widely used in cities throughout Europe (3). As mentioned above, in *The Gutenberg Galaxy* Marshall McLuhan writes that "technologies are not simply inventions which people employ but are the means by which people are re-invented." It is commonly accepted that the free flow of information between social classes that was generated by the Gutenberg printing press was one of the key factors in the religious and sociopolitical changes in European society. It expanded the collective memory of the people, and their capacity to create and share new information with others. From a humble beginning, the movable type printing press permeated the entirety of Europe, sustaining scientific and economic growth. A similar set of immeasurable ramifications would emerge nearly five centuries later with the invention of the internet. Through the invention of the movable type printing press, books became widely available to all strata of society, and this distribution of knowledge became one of the key instruments of the European political system. (See, above, Figure II.4. The diffusion of the Gutenberg movable type printing press.)

As time went by, the use of Gutenberg's printing press was reduced because of the advent of hot metal typesetting. All new technology is bound to get improved upon or even replaced by newer technologies, making the existing one either less attractive or obsolete. This is the normal cycle of technological evolution and invention. The same pattern can be seen, for example, in the history of the steam engine.

The Steam Engine

The history of steam power and the origin of the steam engine go back to the first millennium, but because of the lack of reliable information for this period, I will start with the first recognized inventor, Jeronimo de Ayanz y Beaumont, a Spaniard who obtained the first patent for a steam engine in 1606. In 1698, Thomas Savery patented a steam-powered pump that used steam in direct contact with the water being pumped. Also known as a Miner's Friend, it facilitated the extraction of coal from water-filled mines. In 1712, several improvements were made to the Miner's Friend by Thomas Newcomen to increase its efficiency, leading to the steam engine. At that time, coal, the main source of fuel, seemed abundant in England. In 1764, the redesign of the steam engine by James Watt extended its application to run locomotive wagons and public railways. Thus was born a public transportation system, allowing people to commute outside of city limits. Large numbers of passengers could be transported. Steam engines were subsequently installed in furnaces in England, permitting the development of industries other than cotton mills, such as iron mills, paper mills, potteries, and breweries. Steam locomotives dominated railway transportation until they were replaced by electric and diesel locomotives in the early 1900s. The rapid transformation and development of European society (which was transferred to the American colonies) can also be attributed to two other technological innovations in the late 18th century: vaccinations and domestic gas lighting.

Vaccination against Smallpox

Life expectancy at birth in England rose from approximately 35 years in the 17th century to 41 years in the following century.

This change is attributable to the beginning of better sanitation and food, but also to vaccination against smallpox, which was discovered by Edward Jenner in the 18th century. This terrible disease, manifested in disfiguration, blindness, and death, horrified Europeans for centuries. Vaccination put an end to the propagation of smallpox. Consequently, the population of England and Wales grew from four million in the 17th century to 5.5 million in the 18th century.

Domestic Gas Lighting

This growth led to more social gathering and entertainment. The invention of domestic gas lighting facilitated the shift from a repetitive to a more diversified set of activities. Lighting on farms and streets and in homes also provided protection against predators, thieves, and vandals. This invention even modified eating patterns; the dinner meal took on a social aspect in terms of its composition, participants, and atmosphere.

Income Disparity

Since the First Industrial Revolution, business had been booming and social life seemed to have substantially improved, but this was not the case in all classes of society. Economic disparity in terms of purchasing power existed then, either due to natural causes, such as the bad harvests of 1620, 1630, and 1640, or because of economic factors such as low-paying factory work or one's particular social position. In late medieval Europe, particularly England, there were the nobility, the gentlemen, and the masses (tenant farmers, laborers, and craftsmen), all differentiated by political power, land ownership, and education. Thus, not all the goods produced by the industrial revolution could be purchased by everyone. In addition to budgetary constraints, laborers, including children, faced long work hours in difficult physical conditions. Those kinds of abuses would lead to labor disputes and what was known in the 20th century as the Labor Movement. Furthermore, the transfer of the workplace from the home to the factory, for the sake of efficiency, also concentrated capital in the hands of a few and was accompanied by the transfer of ownership of work instruments, such as the spinning wheel.

With this, we can see a red flag on the horizon, warning about the negative effects of technological development, a theme that will be further developed in Chapter V. James Burke and Robert E. Ornstein, in *The Axemaker's Gift*, describing the "axemakers" as those inventors and innovators who shaped our world and minds, state that "each axemaker's gift was so attractive, not evil or ugly. We always come back for more, perhaps unmindful of the latent external costs." As I have defined technology in the preceding chapter as the expression of the will of our species to live, I believe that a culture of greed and a lack of social responsibility, top to bottom, are more appropriate for diagnosing our societal mishaps or misdirection. It follows, then, that it would be unfair to totally link technology with societal mishaps or misdirection. There are plenty of examples, even today, of the overuse or misuse of innovations. Any step in the course of product or service demand in our journey is the manifestation of our collective choice for more and better as soon as possible. Here is the dilemma. On the one hand, each individual behavior of the collective whole seems to fit the definition of instant gratification, the desire to experience pleasure without delay or deferment. On the other hand, delayed gratification may not always be an easy alternative to adopt because of the feeling that we may not be around to reap and enjoy the novelties currently available.

The Beginning of Mass Production

The mechanization of textile production in a factory environment since the 17th century signaled the beginning of the First Industrial Revolution. Starting in England, it spread to Western Europe and to America by the North Atlantic. A substantial improvement in the standard of living in these societies occurred, at least in the upper class. The 18th century is also called the Age of Reason, with the influences of Bacon, Kepler, and Newton in England, and Voltaire,

Rousseau, and the French Revolution in France paving the way to the Second Industrial Revolution in the middle of the 19th century.

We can fairly say that technology has created the middle class. Access to trades facilitated upward movement in the social pyramid, allowing more people to pursue their dreams through their own efforts—at least in theory. The middle class in the 19th century, facilitated by the technological revolution, was deeply rooted in true individualism, a social theory of economist Friedrich Hayek which posited that the "spontaneous collaboration of free men creates things greater than their individual mind can ever comprehend."

In Great Britain, however, the "British Bee Hive," as portrayed by George Cruikshank (4), showed a certain rigidity between the social classes. As will be seen later, upward mobility in the social pyramid was perceived to be easier in America. There, the middle class was much more broadly defined and inclusive, with greater accessibility for all professions and businesses. With the steam engine modernizing transportation and the manufacturing sectors, increased immigration, and refinements in the division of labor, the economy was growing at an accelerating rate.

The telegraph became popular mainly after the Great Exhibition of 1843 (which will be discussed in the next chapter) and attracted a lot of attention. Within three years, Great Britain was wired with a network carrying many thousands of messages per week. About half of these messages were related to stock and commodity prices, a further third was other business-related messages, and about one in seven were to do with personal or family matters.

Samuel F.B. Morse invented his telegraph system and developed connection capabilities with the financial assistance of Congress in 1843. On May 24, 1844, Morse was able to send the following message from Washington to his friend Vail in Baltimore: "What hath God wrought." By 1851, through Morse's perseverance, 50 telegraph companies were operating in the United States. Subsequently, fast and effective written interstate communication was a fait accompli, strengthening the bonds between families, friends, and business communities.

Business

It is at the business level that the effect of 19th-century techno-logical development can be most felt up to the present day. If the locomotive and the post office have reduced the communication time between states, this is minimal compared to the speed and cost of information-gathering through the telegraph. The acceler-ated development of this new mode of communication favored the effervescent market needs and, to an even larger extent, the growth of firms eager to maximize cost-efficiency and profit. By 1861, the West Coast market, including California, was effectively accessible. Larger markets required more production, facilitating scale produc-tion, and efficient access to the market called for various modes of channel distribution.

Depending on the product, some firms chose the integrated form by affirming their presence or proximity to their market. Others, because of the specific character of their product, chose distributors or what we call today "dealers," because of the required after-sales services. In both cases, the impact of time constraints had been elim-inated. From a marketing perspective, one can easily see the begin-ning of intermediation in the distribution channel from the manu-facturer to the final consumer, or the fully integrated structure of a firm on the national market because of its product specificity. This serves to show how the reduction in time spent upon information gathering accelerates the development of a market and has implica-tions for a firm's business structure as well as its capability to respond to market forces.

Fast forward to today, and the science of marketing has further perfected this elaborate distribution pattern in response to the increased demand of the postwar period. Three major drivers, quite intertwined, have enhanced this flow of goods and services distribu-tion: population growth, the internet, and globalization.

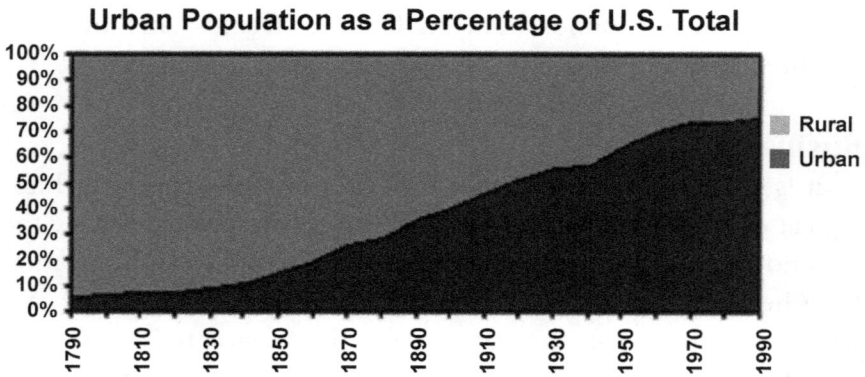

Urban Population as a Percentage of U.S. Total

Figure III.1 Urban population Source: pubs.usgs.gov

Economic Development

As mentioned earlier, the accelerated mechanization of agriculture in the 19th century expanded to industries other than just textiles, consolidating the establishment of the factory system. The application of some of the management principles of F.W. Taylor, such as the division of work, and time management in work rationalization, made factories even more profitable and also rendered unskilled employees more productive. As will be discussed in the next chapter, with this industrialization, the social profile of the United States changed as a result of the influx of immigrants from Europe, Asia, Mexico, and Central America. The total population of American cities grew from approximately 152 thousand in 1790 to 22 million in 1990. Historically, this socioeconomic phenomenon caused by technology and entrepreneurship is the root of the "American Dream," a term coined in 1931 by writer James Truslow Adams. Individual initiative was the key for achieving dreams and a new way of living. America was the land of opportunity because of freedom and the lack of an hierarchical or aristocratic society that determined the ceiling for individual aspirations. This is the essence of the meritocratic philosophy that has sustained the American Dream, albeit with variations over time because of market conditions and technological advancements.

Figure III.2a A mansion in the early 1880s

From Figure III.2a, one can appreciate the quality of life enjoyed by the upper class. The creator of this illustration assumed that electric light bulbs were probably powered by a station in the building's basement. The last decades of the 19th century were satirically dubbed the "Gilded Age" (2) by the American author and humorist Mark Twain because of the extravagant display of opulence resulting from unparalleled technological development. The Gilded Age was also a period of greed, corruption, and horrific labor conditions for women, men, and children. Figure III.2B, showing two young boys working in a textile factory, says it all. It took the introduction of the Fair Labor Standards Act of 1938 to minimize the use of children in the labor force.

Figure III.2b Child labor Source: www.boundless.com (8)

Nevertheless, it was the period when a few important laws were passed, such as the Anti-Trust Act to outlaw monopolies and cartels so as to foster competition, and the Pension Act for Civil War veterans of the Union Army. Regulations concerning utilities, railroads, and industrial working condition were also passed by Congress in the 1890s.

By 1892, there were about 4,000 millionaires in America. But this wealth occurred at a huge social cost (7), leading to the Labor Movement. By 1900, the elite controlled 90% of the United States's wealth. To give significance to this disparity, the economist Joseph Stiglitz in his book *The Great Divide: Unequal Societies and What We Can Do about Them* stated, inter alia, that income inequality at the end of the first decade of the new millennium had never reached the levels of the 1920s.

It must be noted that the need for unskilled workers, although still high, was altered by the new need for semi-skilled and skilled workers. The invention of electricity, the telegraph and, in steel manufacturing, the Bessemer converter can be considered catalysts for this need. As is always the case, diversity in innovation and the concentration of labor create a need not only for general labor but also for specialized workers.

The preceding chapters outlined the most important innovations of the 20th century, but from a societal viewpoint, some inventions have had more impact than others. These innovations included: radio, automobiles, and television in the entertainment sector; insulin, penicillin, and vaccines in the health sector; and programmable computers, transistors, mobile phones, and the internet in the high-technology sector. The last four innovations can be considered the cradle of the digital revolution.

By the middle of the 20th century, home appliances had joined the novelties club, including water and milk sterilization. With these innovations, one can see the introduction of new standards of living totally different from those of the 19th century. The above innovations in the health sector contributed to increased life expectancy: up to 70 years for males and 71.6 for females in 1950.

Figure III.3 Smoke pollution in the English town of Widnes, late 19th century
Source: Wikipedia.org

Although popular, those benefits did not penetrate all social classes until the following two decades. Furthermore, the labor conditions in a production line required continuing attention and physical strength, which took their toll on workers. This was well illustrated by Charlie Chaplin in the silent movie *Modern Times*. Furthermore, the level of pollution caused by the use of coal as a primary source of energy was atrocious. From the 18th to the mid-20th centuries, the resulting smog and soot had devastating effects, not only on the factory workers but also on all of industrialized society.

In London, England, smog killed 700 people in 1873 and 4,000 in 1952. This latter figure precipitated the passing of the Clean Air Act (1956), moving factories and power stations to rural areas. In the United States in late October 1948, air pollution asphyxiated 20 people in Donora, Pennsylvania, and affected more than 7,000 people. Although acid rain was discovered in 1850 the Clean Air Act was passed in the United States in 1970, followed by the Clean Water Act in 1972. Today, environmental issues are still on the table, and legislation is forcing business owners and citizens to be more socially responsible. It is then fair to say that global warming and its consequences are rooted in the First Industrial Revolution. Modern society is painfully learning to be wise in the use of its creations to limit their unintended consequences.

In the 20th century, in terms of lasting impact, considering the spread of electronic devices, the general consensus is that the transistor has been the most important invention. The proliferation of portable radios, cellular phones, and flat-screen TVs would not have occurred had the transistor not replaced vacuum tubes. Miniaturization and the improved price–performance ratio also substantiate this claim. Moreover, the internet and the self-fulfilling Moore's Law prophecy allow more people to own a cellular phone today. Almost five billion people around the globe use a portable phone in one form or another, including smartphones. This is further evidence of price reduction and market penetration of a product over time. According to a recent United Nations report, more families own a cell phone than own a toilet. More important, however, is the increasing role of cellular and internet technology in eliminating illiteracy. According to UNESCO (the United Nations Educational, Scientific and Cultural Organization) in 2015, globally, the youth literacy rate had increased from 83% in 1990 to 90% in 2011, while the number of illiterate youths had declined from 170 million to 126 million.

Subject to the increase in bandwidth in the least developed countries, cellular and internet technology can be seen as an instrument of socioeconomic development in that it raises the level of education and facilitates economic development through global communication.

The statistics about the socioeconomic impact of the internet on the U.S. economy are impressive, and the numbers are increasing at a fast pace. They are subject to variations, depending on the selected metrics. According to Professor John Quelch of Harvard Business School, in the article "Quantifying the Economic Impact of the Internet" (August 17, 2009), each internet job supports approximately 1.54 additional jobs elsewhere in the economy, or roughly two percent of employed Americans.

About 190 million people in the United States spend, on average, 68 hours a month on the internet. A conservative valuation of this time is an estimated $680 billion. The advertising-supported internet creates annual value of $444 billion. With its ever-expanding global communication capacity, the internet is a source of job creation and wealth available to all.

Indeed, the internet has evolved into an instrument for socioeconomic development. It has also changed the education landscape, much like the blackboard did in 1814. It is now well accepted that the internet, cellular phones, and tablets are some of the most effective strategies to eliminate illiteracy and communicate emergency information.

Taking a broader view of the acceleration of technological development, three observations can be made:

1. The benefits of technology are real, but they have not yet permeated all strata of society.

2. Singularity, a concept defined in Chapter II as a period of profound change, will have a soft landing.

3. There may be perils in emerging technology.

1) The Benefits of Technology and Social Strata

It is sometimes a difficult task to accurately assess the social impact of a particular technology, because of the flashing presence of other novelties. Our memory often fails us. For example, few remember black-and-white TV. In the last 20 years, the touch-screen phone replaced the flip phone, which had replaced a brick-sized phone, this latter having replaced a portable valise phone. Closer to us, perhaps still within recent memory, the big desktop computer or slow dial-up internet are now found only in out-of-town surplus stores. Flat-screen 3D smart TVs, smartphones, and tablets are currently part of our daily use or entertainment. In addition, microwaves, programmable ovens and other smart, energy-efficient appliances allow us to make better use of our family quality time. More than 90% of families in developed countries have basic household appliances: refrigerator, stove, washer, and dryer. More than two billion people worldwide own a smartphone. These are examples of new products that achieved market penetration over time. This is not to say, however, that each family owns a refrigerator with an outside ice provider. If I were to depict the haves and have-nots as a pyramid, it would be flattening over time. I also expect that by the next decade, subscription to the internet will be available for a fraction of its current price. Similarly, many items in our near future will be good candidates for price reduction and, considering their utility and convenience, will share the same fate.

Google has promised the commercialization of driverless cars in five years; Google watches and lenses are already commercially available. Recent events suggest that pilotless commercial airplanes and driverless trains could become a reality. Perfected robotic arms in the manufacturing processes, robot assistance in hospital emergency rooms, needle-free syringes, and 3D printing are just a few of the life-changing examples coming to the market.

Technological structural unemployment means jobs lost as a result of automation or other related applications. In agriculture, tractors have replaced horses, and in the next few decades, open-field farming may be replaced by more efficient indoor, environmentally controlled farming, substantially reducing the agricultural workforce. In the price reduction context, the case of vanilla is worth mentioning. Jeremy Rifkin, in an article entitled "The End of Work," reported that Escagenetics, a California biotechnology company, could sell on the global market its genetically engineered version of vanilla for less than $25 a pound, compared to $1,200 a pound for the traditionally produced product; this may leave the economy of smaller vanilla-producing countries in a shambles. Although this event can be characterized as one of the impacts of globalization, the fact remains that at the micro level, a number of vanilla growers will need to find other work. This is not an isolated case and can conceivably be extended to other crops requiring intensive labor, such as rice, corn, and sugarcane.

Taking a look at the car manufacturing sector, the production line of Henry Ford was modified decades ago but recently more substantially by the introduction of robotic arms to eliminate repetitive tasks. Car manufacturers are automating and reducing their workforce to remain competitive on the global market. The same situation prevails in other car-related industries, such as steel and tires. Profit increases by reducing operation costs and/or increasing labor and capital to meet demand.

Nanomedicine

In nanomedicine, the application of nanotechnology to medicine, the promises are immeasurable—from drug delivery for the treatment of cancer, to blood purification and tissue engineering. Nanomedicine is already a multibillion-dollar industry, encompassing a network of more than 200 nanotech drug-related companies.

Molecular imaging & therapy

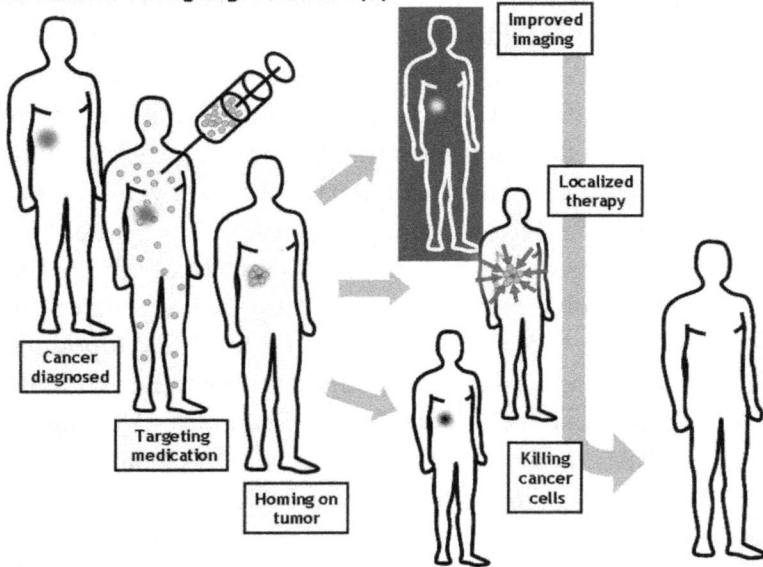

Figure III.4 Molecular imaging and therapy – schematic illustration showing how nanoparticles or other cancer drugs might be used to treat cancer. Source: Wikipedia.org

Table III.1 compares the current disease statistics and the promises of nanotechnology. We are lucky to be living in this period of humanity. There is hope, and I believe that human dignity is at the highest level it has ever been.

Table III.1 Promises of nanotechnology

Facts	Promises
Neurodegenerative diseases affect over 25 million people worldwide.	Therapeutic nanotechnology drugs will re-construct brain cells and restore cognitive function.
Over 322,000 people around the world die from burn-related injuries every year.	Nanopolymer materials will regrow skin and tissue.
521,000 women died of breast cancer in 2012 worldwide.	Nanosensors 1,000 times better than mammograms will be able to find breast cancer, and nano-instruments will be able to destroy and prevent it.
Four of the top 10 causes of death in poor countries are infectious diseases.	Nanotherapies can self-adjust to fight bacterial, viral, and infectious diseases.
382 million people worldwide suffer from diabetes, including 31 million in the Caribbean.	Implanted biosensors can manage glucose and vital signs, releasing medication and detecting emergencies.

But there is another side to the coin. The role of the infinitely small could also be infinitely big in a negative way. There are many books and papers outlining the potential dangers of nanotechnology. Martin Reese, in his book *Our Final Century: The 50/50 Threat to Humanity's Survival*, gives the example of nanosized robots (nanobots) that could be directed to perform harmful acts against humanity—such as injecting toxins or, worse, destroying water—thereby making life impossible on this planet. Some of these issues have already been addressed in a protocol to not allow runaway nanobots (also called "Gray Goo") to replicate without strict controls. The risks pertaining to emerging technology will be discussed in fuller detail in the next few pages.

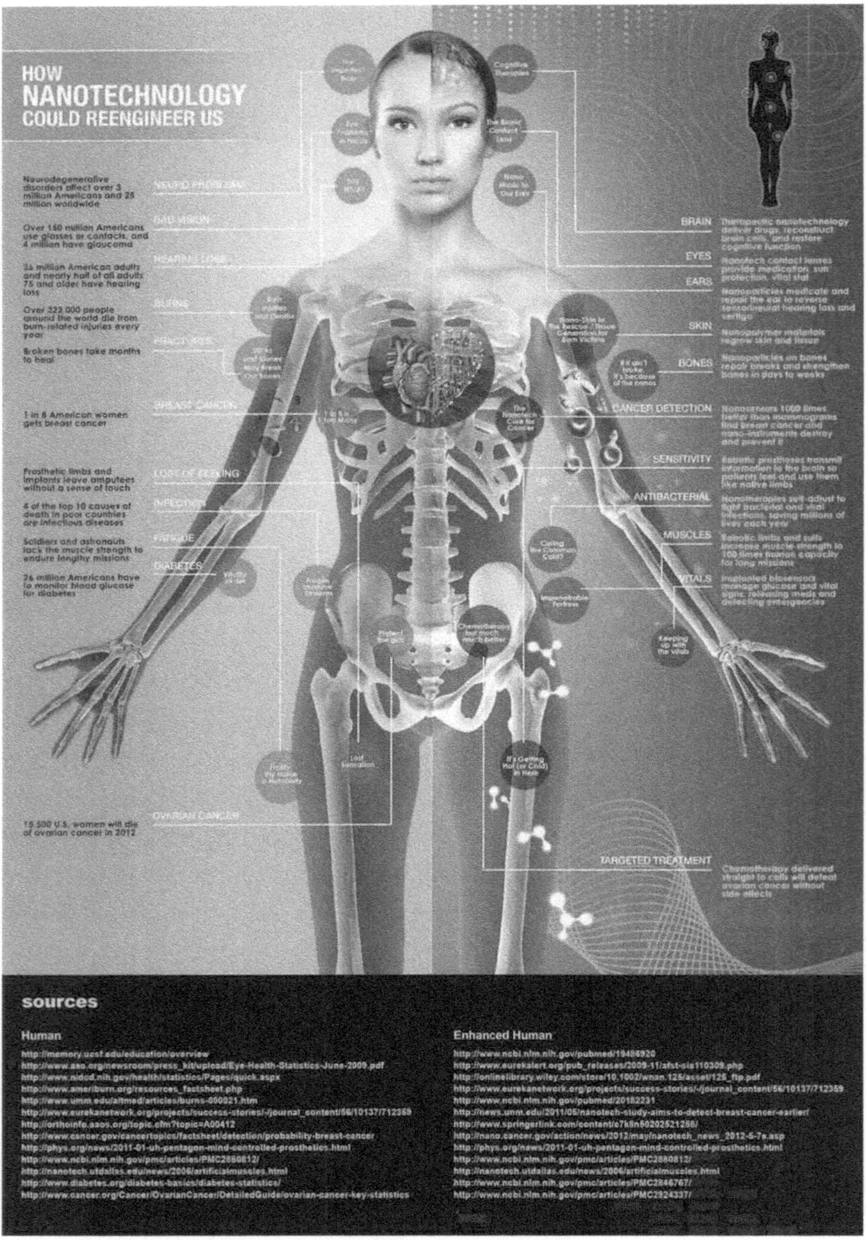

Figure III.5 Potential benefits of nanotechnology – health sector

According to the Association of Clinical Oncology, by 2030 cancer will be the number one killer in the United States, replacing heart disease. People will have more and more new devices implanted into their bodies as nanotechnology continues to penetrate the market.

The growth of nanotechnology

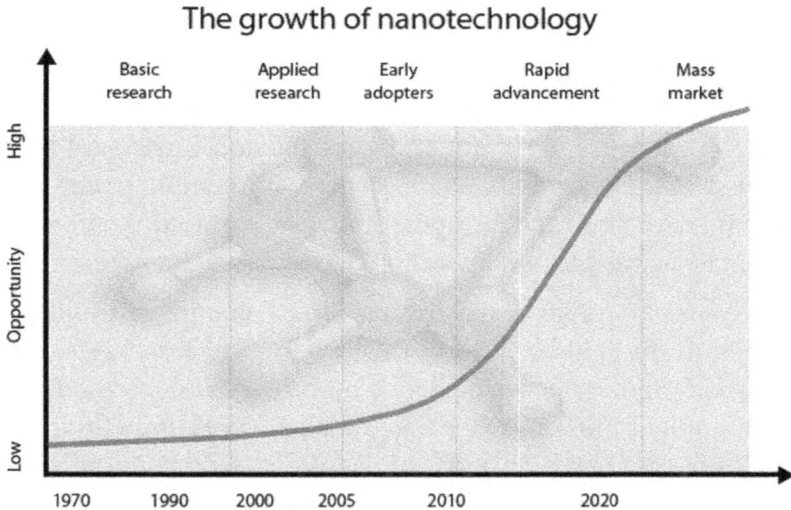

Figure III.6 Growth of nanotechnology
Source: http://www.thatsreallypossible.com/news/1145/

I have inserted Figure III.6, which shows the rapid growth of nanotechnology, right after outlining the potential benefits of nanotechnology for a specific reason. In spite of its potential benefits for humankind, nanotechnology, as in all technologies, also contains the seeds of our destruction. I repeat this message throughout the book to outline the necessity of practical wisdom in all future applications of this technology.

The Need for Skilled Employees

"One machine can do the work of fifty ordinary men. No machine can do the work of one extraordinary man." —Elbert Hubbard

What must be understood is that technological progress leads to innovation, which in turn drives consumption. A country's economic performance is intimately linked to its innovative capabilities, which allow it to remain competitive in the global market. I believe that most of the sensitivity about structural unemployment resulting from technological development revolves around the new skills required at a strategic and operational level of production.

In Hubbard's quote, "extraordinary man" is intended to mean a person with the skills needed in the new high-technology environment that cannot be performed by a machine. In an open economy, an entrepreneur must also face the importation of competitive products. As time goes by, highly skilled people will be the most preferred employees for an organization to boost its productivity in order to remain competitive. It seems, then, that at the national level, a three-pronged approach is necessary to address this required workforce adjustment: retraining to develop those employees with good potential in their current workplace; retraining for other positions elsewhere; and retirement for those who cannot be retrained and who are close to retirement.

In all cases, these initiatives must be reinforced by other financial support policies for the affected employees. At first glance, this approach may appear costly, but the politico-economic cost will be higher if nothing is done. From this perspective, the universal basic income currently under evaluation in Europe may become a necessity driven by automation.

From the various studies on the matter, including the *European Technological Review 2020* by the International Labor Organization, an "equitable distribution of economic gains resulting from improvement in productivity between labor and capital is key to maintaining a proper balance" in the system. Productivity being one of the outcomes of technological progress, the latter is part of the solution rather than the source of the problem. I do not believe that we are heading more and more towards an economic slump or severe social dislocation. But a shift is occurring in the labor market, particularly in the baby boomers' last wave, in which surplus and untrainable employees are affected. Certainly, a labor shift favoring skilled employees is taking place, and enough training and retraining will need to take place to maintain cohesion in our society.

Connectivity

In the context of social benefits, not all benefits are measurable. A new concept has emerged and is permeating society at large: connectivity. New ways of connecting people are among the most revolutionary innovations of the 21st century, with substantial social and political implications.

Information technology, in addition to the myriad quantifiable commercial benefits, maintains family unity. In the not-too-distant past, the breadwinner could spend a considerable amount of time not seeing his/her family. The internet and VOIP (Voice Over Internet Protocol), Facebook, Twitter, LinkedIn, and other types of internet communication have reversed this situation. For a monthly fee of a few dollars, the derived benefits of these communication modes are immeasurable.

But as is often the case, most technologies can also be socially disruptive. Those of us who belong to the first baby-boomer generation can remember the deep-seated closeness of friends and family gatherings. However, the connectedness of these social events has shifted to a different level. People are texting each other more and talking less. At the beginning of this century, Ursula K. Le Guin in *The Wave in the Mind* brilliantly described the majesty of verbal communication in a way still pertinent today: "Speech connects us so immediately and vitally because it is a physical, bodily process, to begin with. Not a mental or spiritual one, wherever it may end."

With the advent of social media, it has become rare for someone at a party not to text on his/her cellul phone, simply maintaining contact with a friend whom very likely he/she has never met in person. Sherry Turkle, director of the MIT Initiatives on Technology and Self, has also substantially addressed this social behavior in *Alone Together: Why We Expect More from Technology and Less from Each Other* and subsequently in *Reclaiming Conversation*.

Digital connection has replaced conversation, and this communication mode gives the illusion of companionship without the demands of friendship. At the root of this profound shift in human behavior, from teenagers to seniors, is this insatiable appetite for immediacy of information from or to the largest number of acquaintances deemed to be friends. As Aristotle maintained, humans are social animals. However, we still should be aware of where this desperate thrust for conviviality may take us. Some of us have probably heard the misadventure of that multilevel marketing agent who reserved a large dining room to recruit some of his social media friends and thereby increase his clientele base. Only ten out of the hundreds of those friends came and expressed some real interest in his business. So much for the claimed friendships established via social media!

At the educational level, the popularity of the internet, laptops, smartphones, and tablets has changed the face of education. They have become instruments in the democratization of education around the globe. The user-friendly character of these gizmos, coupled with a better price–performance ratio, has improved the literacy of millions of people.

At the political level, from the demise of the Soviet Union to, more recently, the events of the Arab Spring, the current state of information technology makes the survival of totalitarian regimes extremely difficult. Facebook- and Twitter-enabled cell phones propagate information outside of the event boundaries, forcing more transparency in the political process. As Abraham Lincoln once said, "You can fool all the people some of the time, and some of the people all the time, but you cannot fool all the people all the time."

As a result of the extensive use mentioned above, the number of internet users, as per the source Internet World Statistics, is impressive. For a technology created three decades ago, its global penetration level is impressive: as of December 31, 2014, this was around 42%, which likely exceeded 50% by the end of 2016.

Internet Users in the World
Distribution by World Regions - 2014 Q4

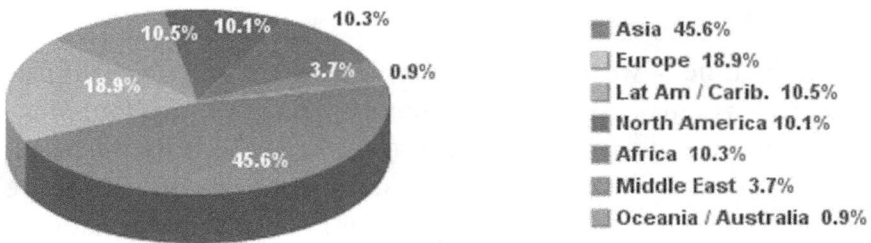

Asia 45.6%
Europe 18.9%
Lat Am / Carib. 10.5%
North America 10.1%
Africa 10.3%
Middle East 3.7%
Oceania / Australia 0.9%

Source: Internet World Stats - www.internetworldstats.com/stats.htm
Basis: 3,079,339,857 Internet users on Dec 31, 2014
Copyright © 2015, Miniwatts Marketing Group

Figure III.7 Distribution of internet users in the world

It would be unwise to underestimate the power of this communication instrument. The impact is outstanding for anyone—for example, business people and political leaders—as part of their communication strategy. Nonetheless, the internet has given an equal chance to everyone to make his/her voice heard, overcoming the impediments of geographical space and of time.

The next 10 years will be very challenging for nurturing democracy, at least with respect to what is expected from developed countries on a number of issues, such as transparency and openness on the one hand, and ensuring homeland security and civil liberties on the other. In theory, these two categories of issues are not mutually exclusive. The growth of the internet, leading to the expansion of a virtual world, may facilitate confusion between the idea or appearance of democracy and democracy itself. We, sometimes unknowingly make accessible our personal information through our search for information, or in our response to an advertisement, thereby revealing our profile in terms of interests, preferences, habits, and finances. The two biggest questions are who owns our personal information and how will it be used? The answer to the first question is the internet service provider; the second is practically unknown. No one knows how your information can and will be used. This issue is still being debated; as strange as it appears, this is the reality. The answer to the second question has raised many issues from the point of view of privacy and civil liberties.

At the time of writing, it has become public knowledge that private citizens' emails and conversations are being recorded for potential future use by law enforcement agencies for security purposes. National security and civil liberties seem to be at odds. It can be argued that in our society, this approach is not viable in that national security should not be maintained at the expense of civil liberties. A solution is surely needed in our democratic society, involving the cooperation of all parties. This issue will be revisited in Chapter VI.

2) Singularity Will Have a Soft Landing

Ray Kurzweil, quoting the mathematician John von Newman (*Singularity is Near*, p.10)—"the ever-accelerating progress of technology ... gives the appearance of approaching some essential singularity in the history of the race beyond which human affairs, as we know them could not continue"—made the observation that human progress advances by multiple constants. We have seen evidence of this accelerating development in the preceding chapter, especially from the 18th century onwards. Another observation that he made is about a particular point in time, 2045, when the compounded effects of technological progress will explode, yielding an era in which human affairs are unrecognizable. Like black holes, no laws in science currently known can explain this situation.

In the same vein, no one at this time can explain what happened before the Big Bang. This is the essence of singularity. In other words, our way of living will be so different that it cannot be explained. Others posit that this new era may cause a rupture in our society or a break in human evolution. Since all the known laws of science fail at the singularity point, of course no explanation can be provided about what is unknown.

I suggest that the singularity will have a soft landing. I make this statement not to calm the technophobes and Luddites but based on the evolutionary behavior of human beings within their environment. I certainly expect a counterargument that the past cannot be used to project about the future, particularly in view of the unique character of the forthcoming technological explosion that has been fermenting over the last two centuries or more. The complexity of human affairs and the resilience of human beings make it risky to draw any conclusion about the future based on a mathematical formula. In addition, "rupture in society" is a strong phrase that may raise anxiety about the future, more specifically regarding those potential events about which we don't know and, for that matter, technology in general. Francis Heylighen, PhD, research professor at the Free University of Brussels, in an article entitled "Socio-Technological Singularity," has addressed the singularity concept and the risk of interpreting it too literally (see Appendix I). I suggest that the singularity will occur as "black swan" events of increasing frequency as time goes on.

It is useful to succintly recount the history of the black swan concept. It was believed that swans have only white feathers. This belief can be traced back to the early 2nd century AD in the writings of the Roman poet Decimus Junius Juvenalis, known in English as Juvenal. A "black swan" was referred to as a bird that did not exist. Until the 17th century, it was common practice to use "black swan" to indicate something's impossibility. In 1697, the unpredictable happened when a team of Dutch explorers discovered some black swans in

Western Australia. This discovery eradicated the belief that swans can only have white feathers. Black swan then became a metaphor to connote the idea that a perceived impossibility might later be disproven. In 2007, the statistician and risk analyst Nassim Nicholas Taleb defined black swan events as rare, unpredictable events having a major impact and inappropriately rationalized after the fact with the benefit of hindsight. Taleb is credited for having extended his concept to all areas of human activities, including the principle of a "black swan robust" society. This is the idea that society gets stronger if, in a nutshell, it acknowledges reality and avoids certain obvious pitfalls—and if it embraces "expect the unexpected" as a normal course of life.

Back to the singularity concept, it is not the first time in our history that major changes with substantial impact have taken place in our society. I can cite many examples, starting with the manifestation of language some 50,000 years ago, the Gutenberg movable type printing press in 1450, the growth of the human population, the internet; the list can go on and on. Homo sapiens is uniquely gifted with language. Whether this phenomenon occurred due to a genetic mutation or was caused by other factors is still a matter of debate.

In terms of social change, one can say that Gutenberg's invention facilitated changes in denominational religions, mainly Protestantism, and in social classes; the same can be said, later on, about the new paradigms developed by Galileo Galilei, Isaac Newton, and others. In other words, with the wisdom of time, we have lived with new paradigms caused by the evolution of science.

I will reinforce my point that the singularity will have a soft landing using two examples: the internet and lifespan extension. We are currently going through a "singularity" in those two areas, and the changes are most notable in government and the insurance industry, the latter because of its vested interest in these areas. The rest of us continue to live our better lives as usual.

Internet

The connectivity created by the internet in reality has profoundly affected our lives. Today, around the globe, no one can imagine life without it, and countless innovations have surfaced because of the web—for example, Twitter, Facebook, iTunes, Skype, to name a few. In *The New Digital Age: Reshaping the Future of People, Nations and Business*, a *New York Times* bestseller, Eric Schmidt states in the introductory section that "the internet is among the few things that humans have built that they don't truly understand... It is at once intangible and in a constant state of mutation, growing larger and more complex with each passing second." Who knows the future of the internet? Since its introduction in the mid-1980s, it has become an indispensable tool of our daily life, a taken-for-granted instrument of education, globalization, and social cohesion. Perhaps some people develop bad social habits and overuse the internet—just as some watch too much television. But since the advent of the internet, our world has become a different place, albeit not drastically.

Life Expectancy

For the sake of clarity, let's differentiate life expectancy from lifespan. Life expectancy, at birth or any given age, is the number of years a person is expected to live based on the statistical average. Lifespan, in contrast, is the length of time a person lives. Both are important measurements of a society's health.

The first wave of baby boomers reached 65 in 2011. In Canada and the United States, this group represents approximately 30% of the population. In the past 60 years, longevity in developed countries has been increasing by a year every four years, which means 15 more years of life over the past 60-year period.

Life expectancy at birth in 2015 in Canada and the United States reached 81.3 and 78.7, respectively. The emergence of an increasing number of centenarians, and perhaps, bi-centenarians, is possible by the turn of the current century. According to the U.S. National Institute of Aging (NIA), the global number of centenarians is projected to increase tenfold between 2010 and 2050. Other demographers are more conservative, putting the average life span at 88 at the end of the current century. In Latin America and the Caribbean, life expectancy at birth rose from 29 in 1900 to 74 in 2010 (subject to regional differences). Increased longevity is also noticeable in the Americas; the 100 million people over 60 years of age in 2006 are projected to double in number by 2020. Nobel Laureate Robert Fogel has defined techno-physio evolution as "the synergism between rapid technological change and the improvement in physiology." Longevity could be considered one of the consequences of this phenomenon.

Lifespan extension is an important event with unknown ramifications that can potentially alter the fabric of our society. Yet this major social phenomenon is occurring unnoticed.

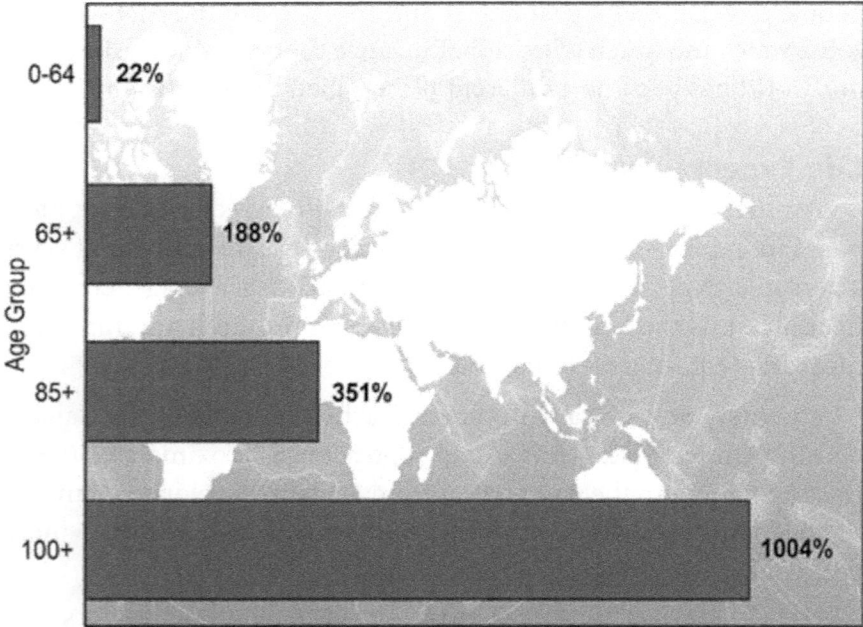

Figure III.8 Percentage change in the world's population by age: 2010–2050
Source: United Nations, World Population Prospects: The 2010 Revision (http://esa.un.org/unp)

Figures III.9a and III.9b are more specific. Together, from a demographic viewpoint, they project by the end of the next decade a leveling off of the population and continuing growth in the number of people over 65.

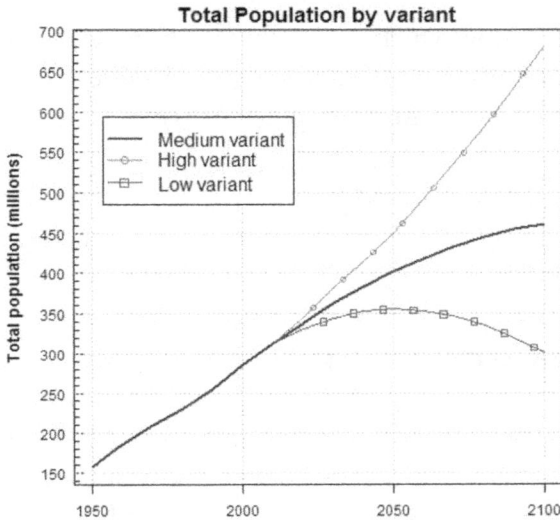

Figure III.9a Total population by variant
Source: World Population Prospects: The 2012 Revision

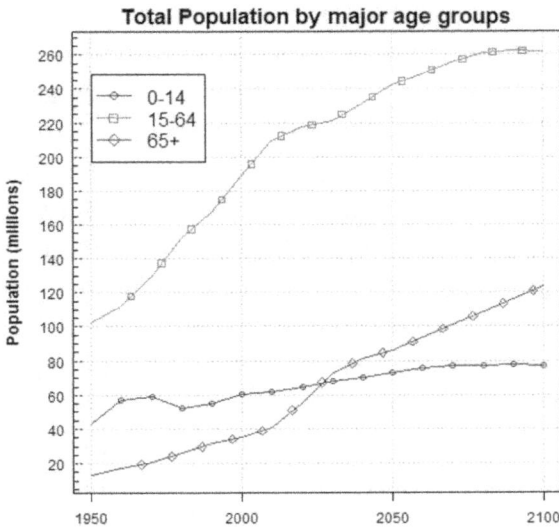

Figure III.9b Total population by major age groups
Source: http://esa.un.org/unpd/wpp/Demographic-Profiles/index.shtm

In our society, senior citizens are increasingly more accepted. Just consider the proliferation of regulated senior citizen homes, private clubs, and in some retail stores. Even the appellation of older people has changed; in Canada, in seniors' newspapers, for example, a 70-year-old person is called a "70-year-young person." This is another smooth transformation which goes unnoticed. Better sanitation, better and more abundant food, and better medical technology are contributing factors to this demographic change.

Again, all we can say is that in the next two to three decades, the world will witness perturbations caused by black swan events of increasing frequency, and hopefully some of those events will be more positive than negative.

3) Emerging Technology Contains Potential Perils

Emerging technology and its rapid development—encapsulated in BRAIN, a multidisciplinary concept including biotechnology, robotics, artificial intelligence, informatics, and nanotechnology—has created some concern. Since worries have been raised by many informed people, they deserve to be considered. The human-made risk detailed at the core of these concerns, particularly in the fields of artificial superintelligence (ASI) and nanotechnology, is that we are dealing with self-learning/improving entities which will be smarter than us and that may drive our civilization from partial alienation to total extinction, despite planned control measures. History has shown that when a superior civilization interacted with a less advanced one, the latter was always absorbed by the former. The real fear is that machines equipped with artificial intelligence, given their accelerated learning curve, may transcend us to a point where we will become their slaves. It is felt that we are unable to confidently know the outcome of any algorithm design to protect us. This seems to imply that Isaac Asimov's laws of robotics, presented in Appendix I, need to be revisited to rule out any unfortunate consequences. Indeed, within these laws and the myriad situations that can emerge in our relationships with a robot, contradiction and misunderstanding may paralyze us. Among the most vibrant opinions expressed are those of:

- Sanders Anderson ("The Five Biggest Threats to Human Existence")
- Bill Joy ("Why the Future Does not Need Us")
- Stephen Hawking ("Artificial Intelligence Could Kill Us All")
- The Global Challenges Foundation ("Synthetic Biology, Nanotechnology and Artificial Intelligence are Three of the Emerging Risks that Threaten Human Civilization")
- James Gips, ("Towards Ethical Robots")
- James Barrat, *Artificial Intelligence and the End of the Human Era – Our Final Invention*

The inventor and futurist Ray Kurzweil has addressed some of these concerns in *The Singularity is Near: When Humans Transcend Biology* (pp. 391–487). There is an absolute recognition that AI must remain "friendly to biological humanity and support its values," but here is the challenge: these machines will think differently than us or, put another way, we may sometimes observe them making unexpected and undesirably "weird" choices, since their analytical capabilities are more sophisticated than ours. After all, we have been building robots for more precision, refinement, robustness, efficiency, and effectiveness to do or make things that we can't do or make.

Going back to my definition of technology, which is the expression of the will of a species to live by taking the form of an ally in our current time, the outcome of a fully developed ASI potentially capable of making humanity extinct would be an awful contradiction. Our conception of intelligence needs to be better articulated to avoid a catastrophe. The fact that there is no turning back makes this matter even more dangerous. I suspect that no substantial breakthrough will manifest itself until former President Obama's Brain Activity Mapping (BAM) project is successfully completed by 2030. Theoretically speaking, it appears that we have time to give thoughtful consideration to ASI, because a lot more needs to be known about the operational functioning of the billions of neurons in the brain, let alone human consciousness and feelings. We are just beginning to discover and understand laws, correlations, and causalities between things which have always existed in our thinking processes. As the development of nanotechnology and computer science allow us to enter into the intimate nature of things, more discoveries are bound to surface. But I would also argue that we have a limited amount of time, given the accelerating development of the self-learning aspect of this technology. Artificial intelligence is already assisting humanity in many areas of life, but the danger is in going too fast and becoming vulnerable to our own creation. A conflict between the created and the creator, as portrayed in the movie *I, Robot*, would be disastrous.

I cannot emphasize enough the need for wisdom in the development of these technologies. Moderation is always better. Without it, humanity may not only not progress but either regress or even be suppressed.

Suggested Readings

(1) Lewis Mumford, "Technics and Civilization"
http://www.realtechsupport.org/UB/MCC/markups2015/
Mumford1_1963_markup.pdf

(2) "Overview of the Gilded Age," *Digital History*
http://www.digitalhistory.uh.edu/era.cfm?eraid=9

(3) Jeremiah Dittmar, "Information Technology and Economic Change: The Impact of the Printing Press"
http://voxeu.org/article/information-technology-and-economic-change-impact-printing-press

(4) George Cruicshank, *The British Beehive*
http://collections.vam.ac.uk/item/O155895/the-british-bee-hive-print-cruikshank-george/

(5) Kathryn Hughes, "The Middle Classes: Etiquette and Upward Mobility"

http://www.bl.uk/romantics-and-victorians/articles/the-middle-classes-etiquette-and-upward-mobility

(6) "Medieval Technology"
http://www.sjsu.edu/people/patricia.backer/history/middle.htm

(7) Working Conditions 1875–1925
http://claver.gprep.org/sjochs/labor.htm
https://www.policyalternatives.ca/publications/monitor/back-good-old-days
http://sageamericanhistory.net/gildedage/topics/capital_labor_immigration.html
Working conditions of Canadian Laborers (1890–1939)
http://workconditions.weebly.com/

(8) Child Labor during the Industrial Revolution
https://www.boundless.com/u-s-history/textbooks/boundless-u-s-history-textbook/the-gilded-age-1870-1900-20/labor-and-domestic-tensions-162/child-labor-870-8841/

CHAPTER IV

Commercialization

"The reciprocal relationships that people voluntarily establish, channel self-interest to mutual advantage and promote a prosperous social order."
— Adam Smith

A great exhibition took place in Hyde Park, London, in 1851. Called the Great Exhibition of 1851, it was the first international display of technological innovations. Held in the Crystal Palace, specifically built for this purpose, the exhibition was the most grandiose and complete display of the finest goods manufactured by humankind. People crossed continents to see this world event, housed in a building made of cast iron and millions of feet of glass.

From the opening on May 1, 1851, to the closing on October 15 of the same year, approximately six million people visited the 1,300 exhibits. The record shows an average of 43,000 visitors daily, with a peak of 109,915 visitors on October 7. This was quite a success. Not until 16 years later, at the International Exhibition in Paris, would this attendance record be exceeded.

At this time, America was undergoing a great transformation. The South, an agrarian society still under a semi-feudal system, was socially different than the industrialized North. The latter was more heavily populated, mostly because of the influx of European immigrants, and was more urbanized and economically diverse. In the South, the combined effects of slavery and fewer and smaller cities were not conducive to technological innovation. This not to say that the South was not wealthy; the skyrocketing price of cotton made investments in diversification less attractive to the wealthy landowners. The American exhibits attracting the most attention were the grain reaper (C. McCormick), the specimens of fire arms (S. Colt), and the rubber goods (C. Goodyear,).

Figure IV.1 The Crystal Palace
Source: https://en.wikipedia.org/wiki/The_Crystal_Palace

Canada, then a colony of Great Britain but on the verge of becoming a self-governing dominion of the British Empire, took this opportunity to show its potential in minerals, timber, and agricultural products. Canada's business goal was to attract investments and stimulate immigration.

Great Britain, starting with the Crystal Palace, which was an architectural wonder on its own, occupied the first place in terms of the number, sophistication, and precision of the machines it displayed.

In his opening speech, Prince Albert, the husband of Queen Victoria, stated that "the aim of the exhibition was to develop a fertile promotion of all branches of human diligence and the strengthening of the bonds of peace among all the nations of the Earth." Great Britain was in a good position technologically, economically, and politically to show its power to the rest of the world. The British Empire ruled over 25% of the world's population, which at that time was 500 million, and 20% of the landmass. As the Scottish writer John Wilson famously said, "The sun never sets on the British Empire." But, ultimately, the British lost hold of this superiority as the combined result of many events. I do not intend to go through the details of the historical events and processes leading to the British Empire's decline. However, among other factors, four events are worth mentioning: the loss of the 13 American colonies, Canada, Australia, and New Zealand as they became self-governing with dominion status; the influence of Adam Smith's capitalist philosophy; the emergence of the United States as a world power; and the politico-economic consequences of World Wars I and II.

As commercialization is the focus of this chapter, we can make a number of observations about the economic success of Great Britain without going into too many details. First, technology, regardless of its source, found a more fertile ground in Great Britain than in other European countries, to a point where it had become the "workshop of the world" (1), that is, the maker and exporter of the vast machinery and equipment needed around the globe. Second, the merchants, later called business owners, and the monarch acted cooperatively in accumulating and protecting the Crown's wealth from its colonies. As an added benefit, Great Britain became a naval superpower. Third, the fact that property rights were deeply entrenched in the Anglo-Saxon legal system strengthened the economic system. The absence or non-maintenance of a legal right of ownership system is an impediment to business development even today in some developing countries. Property rights will be discussed in more detail in Chapter VI. Fourth, the short-term gain of expansionism supported by a mercantilist policy was enormous. This had allowed Great Britain to consolidate its power as a world leader, as shown in Figure IV.2

My purpose in going back to the pre-Second Industrial Revolution period is to show the substantial impact of technology and related external circumstances on society that has extended even to our current time. If commercialization is defined as the introduction to the market of new products or production processes for profit, the list of British innovations from the 17th to the 19th centuries could explain the economic advantages it held. The intensive commercialization of goods and services in Great Britain, as well as the improved living conditions, created an atmosphere of optimism throughout Europe, which crossed the Atlantic and invigorated a young America devastated by the Civil War, in which at least one-quarter of the young labor force was killed. It was then felt in America, as in Great Britain, that technology was the key to a better future. As usual, after a war, many needs had to be met. Liberal professions (engineering, medicine, architecture, etc.) and technical ones (nursing, mechanics, electrical work, etc.) started booming. From 1860 to 1900, 14 million immigrants, mostly Europeans, came to America to improve their living conditions. The necessities of day-to-day life created a pressing demand for all kinds of skills. Individuals felt that prosperity was within their reach because opportunities were offered to all purely on the basis of skills and personal initiatives, at least ideally. This was the beginning of the "meritocracy principle," the genesis of the American Dream, deeply rooted in an egalitarian system open to all. In the last two decades of the 19th century, the American economy grew at a record level, thanks to the commercialization of the many American innovations mentioned in the first chapter.

The refinement of Adam Smith's system of the division of labor and the implementation of Frederic Taylor's rationalization of work activities made American products very competitive between 1870 and 1914, the commonly accepted period of the Second Industrial Revolution. At the business level, the effect of 19th-century technological innovation can still be felt today. If the locomotive and the post office reduced the communication time between states, this was nothing compared to the speed of information exchange via telegraph. The accelerated development of this new mode of communication responded not only to the effervescent needs of the market but, to an even larger extent, to the growth of firms eager to maximize cost-efficiency and profit.

By 1861, the West Coast market, including California, not yet part of the Union, was effectively accessible. A great number of markets led to more production, facilitating a scale economy, and efficient access to those markets created changes in the channels of distribution, which varied depending on the product.

The following graph shows the economic dominance of the United Kingdom resulting from the First and Second Industrial Revolutions, one of the most important periods in the economic history of humankind since the Dark Ages.

Relative Levels of Industrialization, 1750-1900
(U.K in 1900 = 100)

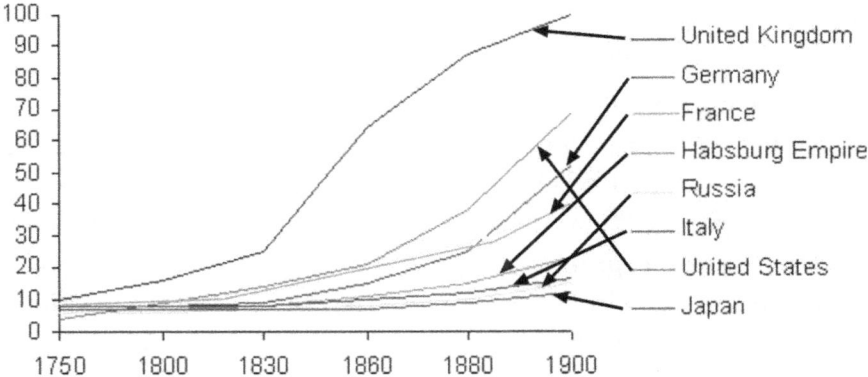

Figure IV.2 Industrialization Source: en.wikipedia.org

In 1840, Great Britain replaced the mercantilist trade policies towards its colonies by a taxation system and removed all trade barriers. Basically, free trade was the new international trade policy.

This move did not prevent Great Britain from maintaining its position as the "workshop of the world" until the beginning of the 20th century. However, production in the United States steadily increased in quality and quantity to such an extent that by the end of the 19th century, American products such as footwear dominated the British market. The price, quality, and style of American shoes made them popular with British consumers and forced British shoemakers to become more competitive (5A).

Trade driven by technological development did not occur solely within the confines of Great Britain. Karl Marx, while living in England, wrote *Das Kapital* (*The Process of Production of Capital – Volume I*), published in 1867. In the introduction, he foresaw Germany following the same industrialization path of Great Britain. His prediction was correct. Not only Germany but also Belgium, France, and the Netherlands joined the industrial parade, capitalizing on the power of the steam engine for production. Germany's pig iron production soared from 40,000 tons in 1825 to 250,000 tons by the early 1850s. Trade, after abandoning mercantilism, substantially evolved between Western European countries, particularly in the agricultural and textile sectors, as a consequence of European population growth (5B).

Commercialization

At this point, society was not far from intensive intercontinental trade, the precursor of globalization. In the 19th century, the collective impact of the availability of choice of products even from outside of the European continent was already apparent. On a more general basis, sometimes it is difficult to really trace the ramifications of an event, be it technological, social, political, or economic. One can suggest that the latent evolution of the necessary components before this new event can emerge remains mostly invisible. In commercialization, globalization may appear to take the marketplace by surprise, but its beginnings can be traced back through millennia, ever since the use of the Silk Road, a network of routes connecting the East to the West circa 114 BC. Globalization can be defined as the process of international integration arising from the interchange of world views, products, ideas, labor, capital, and other aspects of culture. The integration component in this definition has the merit of implying that globalization has its historical roots in the European Age of Exploration, the discovery of the New World, and the international trade that ensued. The chain of events leading to this speedy international exchange of products, services, culture, and people was inevitable. Perhaps as a survival instinct, humans have always had innate curiosity, leading them to search the planet for opportunities and to break barriers in pursuit of greater well-being. Whether this curiosity is a vice or a virtue is a subject for discussion. Suffice to say that in the case of it being vice, there is no immediate or even planned cure for it. If curiosity is not considered a virtue, the course of events from the sedentary society period to the current world would have to be condemned. I simply cannot accept this latter alternative.

The global population rose from one billion in 1800 to 7.2 billion in 2015. A larger population has led to more business exchange around the world. In the last 40 years, we have witnessed increasing global exchange, propelled by greater automated production and mobility of capital. Technological innovations and new production processes are taking place everywhere around the world, making the international market a very sophisticated place to remain competitive. Calibrating a product to meet a specific market may not be as profitable as making it available to multiple markets. One of the key demands of globalization is the ability of an individual or a firm to devise products or services that can meet the needs of markets with varying financial capabilities. Some will succeed, some not. An excellent example of such success is General Electric's Vscan, a portable and easy-to-use ultrasound device now commonly used in rural India, replacing the bulky, expensive, harder-to-use model that GE had previously been offering. The Vscan is now being rolled out in both emerging and developed markets.

In an article about Vscan, authors Navi Radjou, Jaideep Prabhu, and Simone Ahuja use the term "polycentric" to describe the process of creative innovation that led to this product, because, thanks to globalization, the inspiration for it came from "an integrated global research and development team spread across Norway, France, China and the United States." In other words, this fine product is the fruit of the cross-pollination of multiple human endeavors. These authors also use the term "frugal innovation" to describe this approach of creating products and services that are aimed not only at those who cannot afford new technologies but also at high-end consumers, including the many Fortune 500 companies that use them (6).

Earlier in this chapter, I mentioned the inevitability of our current state of globalization in that it reflects the evolution of our society towards greater freedom of choice. Globalization enhances competition for the benefit of consumers. It may provoke the closing of some companies, thereby displacing the resources in favor of more competitive ones. Alan McFarlane, commenting about one of the theories of Montesquieu, wrote in *The Riddle of the Modern World: Of Liberty, Wealth, and Equality* that "if all necessities could be produced within a boundary, this would lead to despotism." Again, it is worth repeating that "commerce encourages freedom. If commerce is the cause of freedom, it is also the consequence of it. Commerce can only flourish where there is a certain freedom." The reciprocal relationship between commerce and freedom suggests that globalization is a noble activity to be nurtured and not discouraged. Commerce infers at least a minimum of communication and respect between trading partners.

The force behind individual determination to achieve unlimited personal ambitions, made possible by equal access to opportunities, was and still is the American Dream. Collectively, this force achieves the greater social good. This was as true in the 19th century as it is today. The parameters (work, skills, market regulations) have changed, but the conceptual framework remains the same. The new game is creativity in original products and services designed for mass consumption and customization, made possible by technologies; this game will continue to be played and will be enhanced by connectivity. Today, what has emerged like a volcano is the contribution of information technology. Globalization, the most efficient evolutionary framework to facilitate the trading of goods and services, is the outcome of a combination of many agreements, providing the opportunity to re-engineer and commercialize new products or services at the international market level.

The internet can be seen as a catalyst of globalization or as an end product of communication instruments. The expansion of the internet as a technology was described in Chapter II, but it is at the international trade level that one can best see its many ramifications. The internet has substantially altered the way companies distribute their products and services. From large corporations to small businesses, increased connectivity has translated into more efficiency in their distribution channels, luring or facilitating certain manufacturers to transact directly with their customers, unlimited by geography or time constraints. This model of distribution, which eliminates a "middle entity" from transactions (a phenomenon known as disintermediation), has permeated most industries. The practicality of the internet as an interface with customers appears attractive from a marketing perspective, and perhaps from a sectorial viewpoint, but not from a corporate approach. Indeed, many companies have neglected to take into consideration the after-sale service factor in terms of the costs and skills required. This is true not only for newly created companies but also for existing ones. In this latter category, some have returned to their previous distribution settings or have contracted with specialized retail stores to service their products. This could be a silver lining for workers who lost their jobs due to disintermediation.

But aside from this challenge, electronic selling transactions translate into substantial savings and the ability to reach the global market in record time; the best example is the dominance of Amazon, which reaches over 100 million people in the e-book industry. Another feature of online purchasing is payment security, ensuring transparency and adherence to terms and conditions of the purchase and sale agreement. Direct e-selling to clients also reflects a growing strategy that saves the manufacturer or service seller millions of dollars, as can be seen in travel, household products and services, and many other industries.

For products or services that can be bought online, the purchaser has a wider choice at the global level, with financial savings or best value for money resulting from increased competition. This trend will include more and more industries as they re-engineer their operations and as the internet continues evolving. On the other side of the ledger, attraction to a product through the internet has physical limits (e.g., gustatory, olfactory, tactile), meaning that with the current state of internet technology, one cannot always taste, smell, or touch the physical product being offered.

As the trend towards disintermediation continues, as information technology evolves and spreads around the world, the benefits will outweigh the costs. From this perspective, internet purchasing will continue to grow because of its current and potential benefits. The greatest benefits yet to come are those associated with the commercialization of the Industrial internet (Ii), in terms of work and materials optimization. Ii can be characterized as the "era of Big Data Exchange." The predecessor of Ii, the Internet of Things (IoT), deals mostly with consumers and business entities' connectivity, whereas Ii connects machines with a processing center, which dispatches the appropriate course of action to be taken. In an example provided by General Electric ("The Power of 1%"), the savings, particularly in the transportation industry, would be substantial. Humanity is stepping into the Fourth Industrial Revolution, following the digital revolution. Currently, some of the parts of an engine/turbine are replaced after a certain number of hours, regardless of whether they seem operational. Linkage with the data provided by built-in sensors would allow the processing center to dispatch the replacement of parts on another basis. The airline industry would then be moving to a precise and accurate efficiency model based on the pre-calculated parts' lifespans. As mentioned earlier, the implications of such a change are profound. This will lead to improved efficiency at the engine manufacturer and maintenance-center level, as more precise and accurate information is fed back to those in program parts management about the economic lifecycle of a part for both the airline company and the manufacturer. In "Industrial Internet: Pushing the Boundaries of Minds and Machines," GE states that it expects to bring huge savings for various components of the industrial sector. For example, a conservative 1% in fuel savings in the commercial aviation industry translates into a potential saving of $30 billion over a 15-year period. In healthcare, a 1% reduction in system inefficiency carries a potential saving of $63 billion over 15 years. This is the Fourth Industrial Revolution, called Industry 4.0 in Germany—Siemens AG being a leader in Industry 4.0, much like GE is a leader in the Industrial internet in the United States.

But these big savings also come with social costs. Are computers taking away jobs? Setting aside the jobs that still cannot be replaced by an algorithm, the short answer is yes. Automation is permeating all sectors of the economy, from factory workers to financial advisors. Earlier, I mentioned the necessity to establish a national retraining program for those able to face the future automation challenge. The well-known economists Dagobert L. Brito and Nobel Prize-winner Robert F. Curl, in "The Rise of Turing Robots Leads to a Fall in Wages," have thoroughly addressed the socioeconomic impact of this forthcoming revolution for developed and developing countries. I foresee an increasing role for the government in terms of transfer payments to maintain a stable society. This is an important issue because structural unemployment can be seen as a source of inequality. Some virtuous citizens feel that they have been abandoned, creating resentment about the unfairness of the economic system.

Changes are difficult, complicated, costly, and sometimes apparently unrealistic to put in motion at the corporate and individual level; think about the re-profiling of American cars in the 1960s to the compact and lean European and Japanese models, or the move of IBM from the house-sized mainframe to the PC. But somehow, the participants in this change must be forgotten.

As time goes on, each invention, innovation, or discovery will continue to happen in a particular period and be nurtured by particular event(s), or will come into existence after a long assembling of its components, spanning years or centuries. In the past, the invention of the car could not have preceded the invention of the wheel, steel, the steam engine, the battery, and electricity. Victor Hugo, one of the greatest poets of the 19th century, once said that "all the forces in the world are not as powerful as an idea whose time has come." Put another way, certain elements need to be in place before certain inventions can be made. This phenomenon was also brilliantly described by Henry Ford:

> "I invented nothing new. I simply assembled the discoveries of other men behind whom were centuries of work. Had I worked fifty or ten or even five years before, I would have failed. So it is with every new thing. Progress happens when all the factors that make for it are ready, and then it is inevitable."

Productivity

It seems common practice to introduce the productivity concept with an analogy or a practical comparison to make the content more digestible. For this purpose, I will use potatoes, a well-known and comestible agricultural product. In Canada, Prince Edward Island (P.E.I.) is one of the largest provincial producers of potatoes. A short review of potato production in this province from 1912 to 2012 has revealed a few interesting points. As shown in Table IV.1, in 2012, the average yield per harvested acre had more than doubled from the level in 1912. Furthermore, the 2012 production level had increased six-fold compared to 1912, while the seeded areas had almost doubled since 1912. I could not undertake a review of the financial gain that the 2012 potato farmers have experienced, because there is a lack of data about the 1912 average wage in P.E.I. In addition, the monthly consumer price index (CPI) data in this sector did not begin until 1914. Nevertheless, all other things being equal, it is reasonable to assume that 2012 farmers have greater purchasing power than their 1912 ancestors. This financial gain happened not because the 2012 farmers were harder working and more robust than a century ago. And a one-acre surface in 1912 is the same as in 2012.

Another example is the production of corn in the United States. In 1933, the average yield was 33 bushels per acre, compared to approximately 147 in 2011.The higher output per acre in each case was the result of developments in disease-resistant seeds, effective planting and harvesting machines, better production processes, and better farm management. In a nutshell, this is productivity or, more specifically, "total factor productivity," which will be covered in the following pages. This explains the growth of agricultural production in Canada and the USA, considering the decrease in the agricultural workforce and the growth of the population in both countries.

In economics, productivity is a measure of gain. Productivity increases when fewer inputs are used in the production of a specific output. One of the many drivers of gain is the technology used in the production process to increase the output while keeping the other inputs constant or at a reduced level.

Table IV.1 Production of potatoes in Prince Edward Island, Canada

	1912	2012
Seeded Area (Acres)	33,000	89,500
Average Yield/ Harvested Area In Tons	6,130	13,875
Production In Tons	202,500	1,239,700

Source: Conversion from Statistics Canada Database

As I am not an economist, I will be talking in general terms, recognizing that I may frustrate economists in this process, for lack of specificity. It is well accepted that productivity depends on a number of factors, such as capital investments, demand, market structure, skills, infrastructure, and competition, to name a few. But before expanding on productivity, we should consider entrepreneurship. This will be discussed later, but for the time being, let's keep it in mind. It denotes the ability to plan, organize, and manage a business entity and its inherent risks.

Productivity is a measure of efficiency in production—in other words, how much output is obtained from a set of inputs or, according to the U.S. Bureau of Labor Statistics (BLS), the real value of output produced by a unit of labor during a specific time. The U.S. economy has gained considerable ground since the beginning of the 20th century and has become the world's economic leader (see Figure IV.3a). We can observe the growth of the population and the fact that the number of working hours is almost the same over the time period covered by this chart. Some capital investments in machinery have been made, which explain the upward trajectory of the graph. Productivity growth in the USA has improved the quality of life of society by making products and services available around the world. However, the productivity growth rate has been declining since 1950 (see Figure IV.3b). Various graphs from different sources and incorporating different variables show the same decreasing productivity trend in the USA and, for that matter, in most developed countries.

Labor productivity in the business sector, first quarter 1947–fourth quarter 2013
Index (2009 = 100)

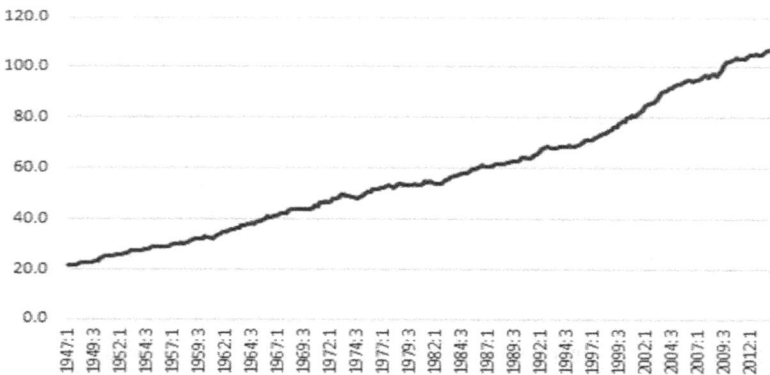

Source: U.S. Bureau of Labor Statistics.

Figure IV.3a Labor productivity in the business sector

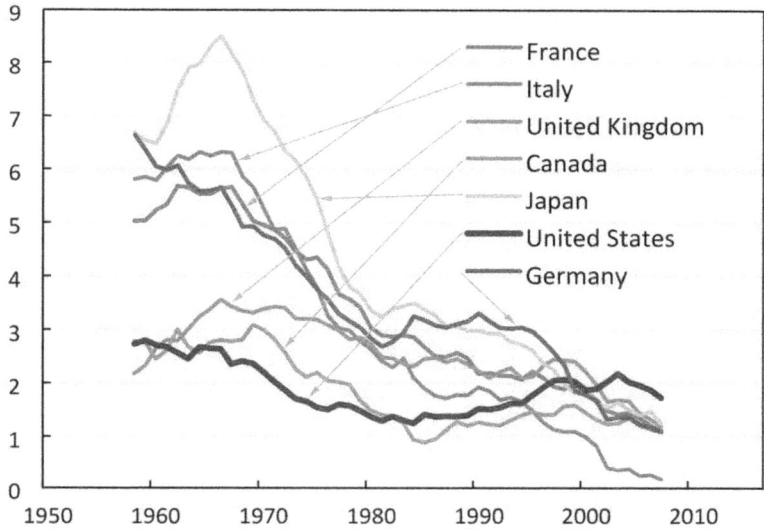

Source: Conference Board; CEA calculations.

Figure IV.3b Productivity in advanced economies
Source: Conference Board Total Economy Database and U.S. Bureau of Labor Statistics –
Nonarm Business Sector

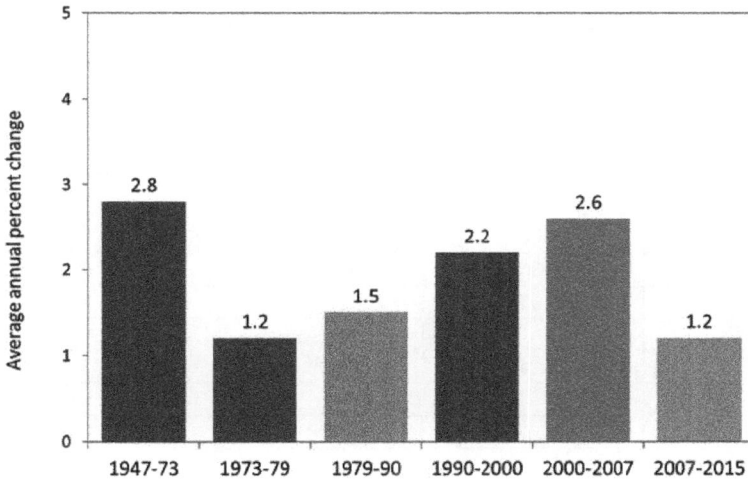

Source: U.S. Bureau of Labor Statistics

*Figure IV.3c Average Productivity Change in the NonFarm Business
Sector, 1947-2015*

It is known that productivity in most developed countries has been on a downward path for the last two decades, excluding the positive effect of information technology between 1990 and 1995 (see Figure IV.3b). It is worth noting that the United States had the second lowest productivity rate at the beginning of the period covered. However, by 2010, the USA dominated the other developed countries. Several major events have affected the global economy; the two-decade recovery period subsequent to World War II, the energy crisis of the 1970s, the non-convertibility of U.S. dollars into gold in 1971, the dot-com boom of the 1980s, and the Great Recession of 2007–2009, among others, are embedded in the graph presented in Figure IV.3c.

Table 1: Improvements in Living Standards, 1870 to 2010

Period	Total Factor Productivity (Average Annual Growth Rate)	Main Sources of Growth	Change in Life Expectancy at Birth (Years per Decade)
1870 to 1900	~ 1.5% to 2%	Transportation, communications, trade, business organization	1.3
1900 to 1920	~ 1%		3.2
1920s	~ 2%	Electricity, internal combustion engines, chemicals, telecommunications	5.6
1930s	~ 3%		3.2
1940s	~ 2.5%		5.3
1950 to 1973	~ 2%	Widespread	1.4
1973 to 1990	< 1%		2.4
1990s	> 1%	Information technology	1.7
2000s	~ 1.5%		1.4
1870 to 2010	~ 1.6% to 1.8%		2.3
1950 to 2010	~ 1.2% to 1.5%		1.8

Sources: Field (2012), Gordon (2010), Carter et al. (2006), Center for Disease Control and Prevention (http://www.cdc.gov/nchs/data/dvs/deaths_2010_release.pdf).

Figure IV.3c U.S. average productivity change in the nonfarm business sector

If productivity trends are important for analyzing and understanding the direction of an economy, it seems that a more inclusive tool is needed, to obtain a complete picture. Total factor productivity (TFP), also called multi-factor productivity, is a residual which accounts for effects in total output not caused by labor and capital. If all inputs are accounted for, this residual can be taken as a measure of an economy's long-term technological change or technological dynamism. Although this approach has long been debated because of the estimating factor, TFP is relevant in the context of this book because it infers that a decrease in productivity is associated with a lack of innovation.

Although other factors should also be taken into consideration Figure IV.3c makes the inference to a lack of innovation reasonable. The peak percentages of 2–3% in this table coincide with the manufacturing of (lasting) innovations leading to high-level economic activities. From 1900 to 1970, in spite of two world wars, inventions, innovations, and consumption were going full tilt and changing people's lifestyles. In the early 1970s, the energy crisis and the non-convertibility of the U.S. dollar into gold in 1971 substantially increased the cost of materials, negatively impacting productivity until the 1990s. I believe that information technology made a substantial contribution to productivity, as reflected in the improved percentages for the first decade of the new millennium, but the impact on the labor force in terms of the required skills made this contribution less obvious.

On the surface, there would appear to be no solution to the current economic challenges facing most developed countries. But not necessarily. Earlier, I mentioned entrepreneurship. In fact, it is the fourth factor of production in the free world economy, the other three being land, labor, and capital. Entrepreneur has been defined in many ways, but the risk element in all of those definitions was recognized in the 18th century by the famous Irish-French economist Cantillon: "a person who pays for a product at a certain price and resells it at an uncertain price." It is defined in almost the same way in many dictionaries: "an owner or manager of a business enterprise who makes money through risk and/or initiative." A businessperson/entrepreneur or investor must have a strong conviction about the financial viability of his or her idea, as well as the mindset and the skills to succeed within the limits of the available resources. All entrepreneurs are motivated by either a perceived opportunity or necessity, the former being the case more often than the latter.

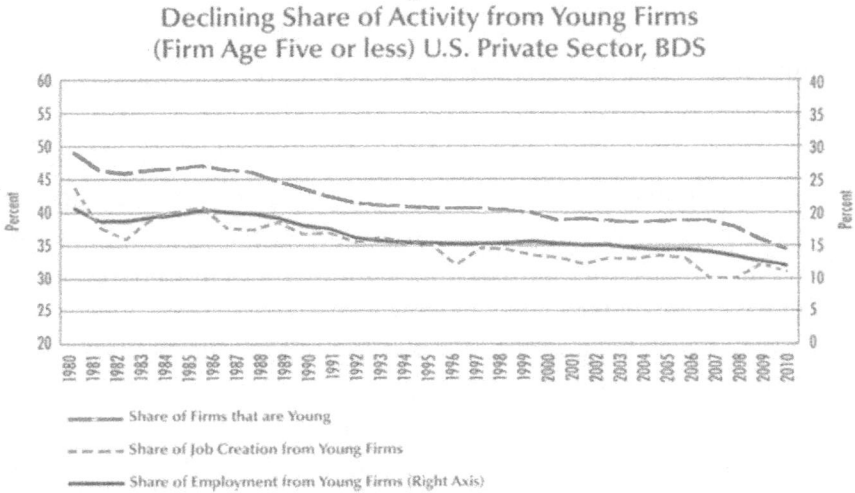

Declining Share of Activity from Young Firms
(Firm Age Five or less) U.S. Private Sector, BDS

— Share of Firms that are Young

- - - Share of Job Creation from Young Firms

— Share of Employment from Young Firms (Right Axis)

Figure IV.4 Declining share of young firms
Reproduced with the permission of Kauffman.org

Not all startup businesses will make it to the big entrepreneur league; this is the hard reality. However, startups can be considered the feeder group of productivity. Some will succeed, but most will fail within the first five years of operation. However, this in-and-out, this elimination process, positively affects productivity by allowing resource reallocation to the fittest survivors and thereby sustains economic growth. As shown in Figure IV.4, from the Ewing Marion Kauffman Foundation, competition is becoming less challenging, in that startups, meaning young firms of five years or fewer, are on a downtrend.

In Figure IV.4, the dashed line representing the proportion of young firms from all the startup firms shows a declining trend since 1980, as does the share of employment of these young firms (the solid line) and their share of job creation (the dotted line). The implication is that the slow pace of productivity growth is, among other factors, related to the lack of entrepreneurship. In other words, the number of firms entering business does not exceed those exiting. Some explain this downward trend by growing risk aversion amongst small-business entrepreneurs. This position cannot be supported, in view of the large pay-off anticipated by business owners for the risks they take. Be that as it may, the slow pace of productivity also affects the resilience of the economy. Indeed, the adaptive capacity of an economy to respond to shocks is dependent, among other factors, on such phenomena as new business entities formation, the availability of appropriate skilled labor, capital, and entrepreneurship.

The U.S. Department of Labor has characterized the 21st century as a new economy powered by technology, fueled by information, and driven by knowledge. The popular belief is that creative ideas driven by the proper mindset and business skills can be financed in myriad forms from the financial network. What then is missing? Why are the number of startups on a downward path? As mentioned earlier, various reasons have been put forward, such as a slowdown in the overall economy, more recently the busting of the dot-com balloon, the Great Recession of 2007, and the scarcity of high-skilled labor. But there is another probable explanation. I believe that it relates to the technological frontier concept, the position that the United States and other developed countries occupy now. The creative economist Giovanni Dosi in his book *Innovation, Organization and Economic Dynamics* defines the technological frontier as the "highest level reached upon a technological path with respect to the relevant technological and economic dimensions." Developing and less developed countries also have their technological frontiers. They progress through imitation of developed countries' production and practices based on their levels of capital and skilled labor.

In developed countries, the current slow pace of the economy appears to result, among other causes, from the fact that the innovative effect created by all the startups using past methods is inadequate to raise the productivity to a higher level. Former U.S. Vice-president Albert Gore's statement about the global economy in *The Future: Six Drivers of Global Change* can appropriately be used in this context: "National policies, regional strategies, and long-accepted economic theories are now irrelevant to the new realities of our new hyper-connected, tightly integrated, highly interactive, and technologically revolutionized economy."

A number of suggestions have been put forward, including the softening of some restrictions on immigration in selected areas because it is recognized that immigrants display a high entrepreneurial spirit. In the same vein, this could also increase the pool of available high-skilled and semi-skilled workers, the lack of who drives jobs abroad. Others suggest more tax incentives for new businesses, a greater role for unions (particularly for people in low-paying jobs), and increased government participation in the redesigning of retraining programs for each particular industry. Taken together, these suggestions may result in a significant change at the low small-business startup level. In addition, it has also been observed that countries at the technological frontier can benefit from more post-graduate recipients as well as from research and development funding. In essence, a different mindset and environment are needed for the creation of radically new ideas, products and services, production processes, and techniques supported by robust business models. Together, they may lead to productivity, competitive prices leading to employment, and economic prosperity,

As can be seen, the components or rather the ingredients of creativity cover many grounds. In 2015, the urban theorist Richard Florida, currently director of the Cities Project at the Martin Prosperity Institute housed at the University of Toronto's Rothman School of Management, co-authored with Charlotta Mellander and Karen King a Global Creativity Index (12). According to these authors, in a proper environment, technology, talent, and tolerance drive economic growth by fostering creativity. They examined at great length the impact of these drivers, abbreviated as the 3Ts, not only separately but also together in 139 countries, covering all economic levels, in order to come to an overall ranking. Table IV.2, which presents data for the top ten countries for the sake of conciseness, shows the ranking of these countries out of 25 for the 3Ts and their overall global creativity index. The full table can be seen in Exhibit 11 of the report, referred to at the end of this chapter.

Rank	Country	Technology	Talent	Tolerance	Global Creativity Index
1	Australia	7	1	4	0.970
2	United States	4	3	11	0.950
3	New Zealand	7	8	3	0.949
4	Canada	13	14	1	0.920
5	Denmark	10	6	13	0.917
5	Finland	5	3	20	0.917
6	Sweden	11	8	10	0.915
7	Iceland	26	2	2	0.913
8	Singapore	7	5	23	0.896
9	Netherlands	20	11	6	0.889
10	Norway	18	12	9	0.883
----	---	---	---	---	---
31	South	Korea	1	50	70

Table IV.2 The Global Creativity Index (GCI)—overall rankings
Source: Social Progress Imperative Reproduced with the permission of the Martin Prosperity Institute.

A few observations can be made:

1. South Korea, ranked number one in technology, is not even close to the top of the list, although technology has always been considered one of the key drivers of progress and wealth.

2. Taken in isolation, diversity and openness to talented people are not sufficient to ensure economic growth through their connection with tolerance. The social factor, in terms of countries (Canada, Iceland, and New Zealand) having the highest level of tolerance, plays an important role in the

ranking; Canada occupies the first position in tolerance in part because of its multiculturalism policy. As Prime Minister Wilfrid Laurier said on June 25, 1901: "Fraternity without absorption, union without fusion." This declaration was amplified 94 years later by President William J. Clinton: "In a world darkened by ethnic conflicts that tear nations apart, Canada stands as a model of how people of different cultures can live and work together in peace, prosperity, and mutual respect." Canada is one of the best countries in which to live because of the culture, compassion, and forbearance reflected in the social programs of this country and the diversity of people working in the creative class; 44% of Canada's population work in the creative class. For the sake of clarification, in the GCI report, the creative class includes workers in science, technology, and engineering; arts, culture, entertainment, and the media; business and management; education; healthcare; and law.

In addition, it is worth noting that most Nordic countries appear in the top ten of the list. The GCI report suggests that these nations combine a high level of competitiveness with a relatively low level of inequality. In the Social Progress Index mentioned in the next chapter, it should not be surprising that these nations also occupy the top ten of that list, although the metrics used for the two reports are different.

Overall, social policies aiming to eliminate inequalities on balance are important factors, not minimizing the fact mentioned above that new ideas, new production processes and techniques, and successful business models leading to competitive prices in the free market are the most crucial strategies to transform the economy. The foregoing was intended to outline the challenge of passing the technological frontier. The U.S. GDP will still continue to increase yearly, but at a moderate pace. It is forecast to reach $19.2 trillion in 2016, a 3% increase over the forecast for the preceding period—an ambitious percentage increase, at least in my view.

While there is a lot of talk about the desirability of sustainable economies, there appears to be less emphasis on the notion of resilience. In fact, a sustainable economy cannot exist if it is not resilient. Among many definitions, a resilient economy includes a proactive and policy-induced system which embeds a capacity to absorb economic shocks from internal and external sources and the robustness to recover at a level equal to or exceeding the pre-shock level. Examples of shocks are the dot-com bust, natural catastrophes, vulnerability to international markets in terms of imports/exports of essential products, and political factors.

Conceivably, the economies of most developed countries are resilient and have mostly recovered from the Great Recession, besides the housing sector in certain geographical locations. But there are developed countries that have not been affected by the Great Recession, and I became interested in what they have or not have done. A case in point is Singapore, whose GDP from 2007 to 2014 is shown in Figure IV.5

The resilience of Singapore's economy is supported by a combination of factors that took place many decades ago. Among those factors I have noted the stability of Singapore's political system but more importantly the role of the government in education and business development. I must also mention that this country, in spite of an appreciable GDP per capita of $78,958.09, has its share of social and economic challenges, such as economic disparity and weak productivity (14).

The above example is not presented as a model for other developed countries that are structured differently. However, developing and less developed countries could benefit from the Singapore experience of long-term planning.

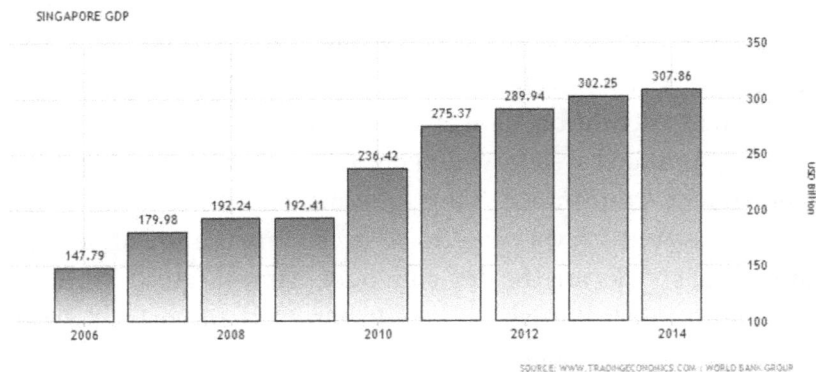

SINGAPORE GDP

Figure IV.5 GDP in Singapore 2007–2014 Source: www.tradingeconomics. com – World Bank Group

People have a natural tendency to overreact to shocking events and more often to attribute to repetitive events the status of irreversible changes. It is unlikely that the current moderate growth will continue to prevail for long, because new paradigms, a new form of growth, will take place, driven by technology and social changes resulting from enhanced education and hopefully the government's intervention to reduce inequalities. As the economist John Maynard Keynes said in his superb book *The General Theory of Employment, Interest and Money,* "The difficulty lies not so much in developing new ideas as in escaping old ones." Indeed a (technological) frontier is liminal, a temporary place, a threshold full of promises and challenges, hopefully more of the former than the latter.

Past this period of modest growth, a new era of abundance is ahead. But are we ready for it? In fact, this question is the title of a TED Talks video, "How Technology Works in the Age of Abundance" (13), which is about the future of economy and society. Cross-pollination of technology around the globe through increased communication will make products and services cheaper and the standard of living higher. The same will be the case for new business owners: the successful ones will move from being unknown to becoming superstars in a short period of time, because information technology will magnify their activities. In an economic growth scenario, the GDP will move above the moderate increases recently experienced. I am not suggesting any precise number but rather predicting a transformed economy, driven by the application of new technology, new business models, and social policy changes.

In this new era, conceivably some people may choose to enjoy life by consuming more products and services or dedicating more time to leisure activities. The leisure society concept could be revived in the next few years as a result of this new economic growth. Suggested since 1931 by John Maynard Keynes in *Economic Possibilities for Our Grandchildren* (3), the leisure society concept has been subject to major variations over the last century (16).The positions of a wide variety of authors on a leisure society can be found in Appendix I. Some scholars have argued that our society is not ready for it or that a leisure society will provoke social instability and conflicts. Others believe that in an abundance context, people will be busy consuming more, or will even become stressed by the many choices they face. As the number of options increases, so does the feeling of being harried. As productivity increases, people will aim to draw fulfilment from their leisure time at least as much as from their time at work. I believe that the start of a leisure society may occur in the next three decades, contrary to previous erroneous speculations, for two main reasons:

1. Robotics will continue to improve manufacturing production processes, simplify household work, and assist in the provision of services even in the liberal professions; yes, lawyers, accountants, financial advisors, physicians, and engineers will not be spared.

2. Mortality and birth rates will continue to decrease. It is inevitable that society, or at least a part of it, will experience increased free time. More likely, new businesses will be set up to advise people on how to use their free time in a more constructive and enjoyable way. I assume that by that time, people will have developed

an appreciation for being alive and the capacity to better manage their needs and wants.

Suggested Readings

(10) Britain, the Workshop of the World
www.bbc.co.uk/history/british/victorians/workshop_of_the_world_01.shtml

(2) The American Dream
https://en.wikipedia.org/wiki/American_Dream

(3) John Maynard Keynes, *Economic Possibilities for Our Grandchildren*
http://www.econ.yale.edu/smith/econ116a/keynes1.pdf

(4) R.L. Martin, "Regional Economic Resilience, Hysteresis and Recessionary Shocks," *Journal of Economic Geography*
http://www.euregionalgrowth.eu/download/Regional%20economic%20resilience,%20hysteresis%20and%20recessionary%20shocks.pdf

(5A) The Effect of the American Export Invasion on the British Boot and Shoe Industry 1885–1914
journals/journal-of-economic-history/article/div-classtitlethe-effect-of-the-american-export-invasion-on-the-british-boot-and-shoe-industry-18851914div/3111B69E5523263DF713A40
5C6C07C5E

(5B) History of Europe. Revolution of the Growth of Industrial Society
https://www.britannica.com/topic/history-of-Europe/Revolution-and-the-growth-of-industrial-society-1789-1914

(6) N.S. Poblador, "Finding a Common Framework for the Analysis of Social and Institutional Change: A Retrospective and an Exploration," *Philippine Science Letters*
http://philsciletters.org/2014/PSL%202014-vol07-no01-p146-154%20Poblador.pdf

(7)Why America Must Embrace Innovation
http://globalpublicsquare.blogs.cnn.com/2011/06/20/america-must-embrace-global-innovation/.)

(8) Shawn Sprague, "What Can Labor Productivity Tell Us about the U.S. Economy?" *Bureau of Labor Statistics* http://www.bls.gov/opub/btn/volume-3/what-can-labor-productivity-tell-us-about-the-us-economy.htm

(9) Why Economic Growth is Getting Harder and What to Do About It. http://www.huffingtonpost.com/brink-lindsey/why-economic-growth-is-ge_b_853347.html

(10) J. Rifkin, *The End of Work: The Decline of the Global Labor Force and the Dawn of the Post-Market Era* https://www.amazon.ca/End-Work-Decline-Global-Post-Market/dp/0874778247

(11) Rose Jacobs, "Humans Need Not Apply: Rise of Robot Factories Leading 'Fourth Industrial Revolution,'" *New Illuminati* http://nexusilluminati.blogspot.ca/2015/03/humans-need-not-apply-rise-of-robot.html

(12) Richard Florida, Charlotta Mellander, and Karen M. King, "Global Creativity Index 2015," *Competitiveness and Prosperity* www.martinprosperity.org/content/the-global-creativity-index-2015/

(13) "How Technology Works in the Age of Abundance" (YouTube) https://www.youtube.com/watch?v=O5dVGFk5aWE

(14) J.L. Martinez, "The Singapore Paradox," *SlideShare* http://www.slideshare.net/jlaguardiamartinez/the-singapore-paradox-41537028

(15) S.B. Linder, "The Harried Leisure Class" http://opus1journal.org/articles/article.asp?docID=145

(16) A.J. Veal, *The Elusive Leisure Society* www.leisuresource.net

CHAPTER V

Progress: Fiction or Reality

"Every day you may make progress. Every step may be fruitful. Yet there will stretch out before you an ever-lengthening, ever-ascending, ever-improving path. You know you will never get to the end of the journey. But this, so far from discouraging, only adds to the joy and glory of the climb."
—Winston Churchill

"The offer of certainty, the offer of complete security, the offer of an impermeable faith that can't give way, is an offer of something not worth having. I want to live my life taking the risk all the time that I don't know anything like enough yet; that I haven't understood enough; that I can't know enough; that I'm always hungrily operating on the margins of a potentially great harvest of future knowledge and wisdom. I wouldn't have it any other way." —Christopher Hitchens

These two opening quotations encapsulate some aspects of progress used in the current context to mean the ascending path of humankind in a long journey towards progress, the potential reward of risk taking while the present and a potential future are unfolding.

But before getting into the very nature of progress and addressing whether it is a fiction or a reality, I feel compelled to use this quote from Bertrand Russell in one of his essays, which points in the direction that I intend to take: "Change is one thing, progress is another. Change is scientific, progress is ethical; change is indubitable, whereas progress is a matter of controversy."

Progress is one of those subjects on which consensus is difficult to reach. The different perceptions arising from each definition often leads to lengthy and incomplete discussions, particularly in the areas of social, economic, technological, and ethical progress. The case can be made, however, that these four areas of progress are intertwined and that, to some extent, one leads to the other.

The invention of tools, the discovery of fire, the emergence of language, the transmission of written information, the way we interact with each other are important steps in our evolution. On a general basis and from a chronological perspective, I believe that most people would agree that we have progressed over the last few centuries. We have fewer wars, and we have mechanisms in place to settle our differences without using armed force. We no longer burn people because of their beliefs. For instance, in 1600, the philosopher and mathematician Bruno Giordano was burnt alive for saying that there is life on other planets. This would be an inconceivable response today. We are also far away from the sordid activity of cat-burning for public amusement that was common in medieval Europe. This practice continued well into the early 19th century; cats were perceived as a source of witchcraft. Surely our societies have become more tolerant. This position is supported in Steven Pinker's acclaimed book *The Better Angels of Our Nature: Why Violence Has Declined* (2011), in which he suggests that there are fewer crimes, fewer atrocities against our fellow beings, and even less poverty than ever before. In fact, according to the World Health Organization, more people die by suicide every year (approximately 800,000) than the combined amount of conflicts, wars, and natural disasters. By legislation, children no longer work in factories. Work safety is now mandatory in all factories. In addition, most countries, to the limit of their financial means, have developed some social programs to assist their less fortunate citizens, in recognition of respect, dignity, and empathy towards others. Taken together, those achievements are indeed good evidence of social progress. However, if the focus is moved to a few peculiar challenges, the answer needs to be more nuanced. For instance, one million people still live in abject poverty and horrible conditions. In addition, at the global level inequality is growing. The gap between the haves and have-nots is deepening, thereby creating social dislocation with strong negative ramifications for democracy. Privacy and, for that matter, civil liberties are at risk as a result of the rise of terrorism. As real corrective actions are being taken, I can continue to place hope in the quote from Sir Winston Churchill mentioned at the beginning of this chapter and believe that social progress is being pursued.

But social progress is also the idea that citizens of societies have the opportunities to sustain their quality of life and reach their full potential. Social progress then becomes related to economic progress.

Gross domestic product (GDP) is commonly used to measure the total output of a country's economic performance in terms of the value of the marketed goods and services for a particular period. The following figures from the World Bank show the world's GDP from 1960 to 2015. One can see a constant progression with very few exceptions, such as the surge in globalization in the 1980s, the impact of the 2008 Great Recession, and the fall in petroleum prices in 2014.

Table V.1 World GDP by year (Dec 31) and value (trillions USD)
Source: World Bank https://ourworldindata.org/economic-growth

Year	Value	Year	Value	Year	Value
2015	73.89	1996	31.40	1977	7.16
2014	78.38	1995	30.73	1976	6.34
2013	76.53	1994	27.69	1975	5.84
2012	74.44	1993	25.79	1974	5.24
2011	72.92	1992	25.34	1973	4.56
2010	65.65	1991	23.85	1972	3.75
2009	59.85	1990	22.52	1971	3.24
2008	63.16	1989	20.01	1970	2.94
2007	57.60	1988	19.07	1969	2.67
2006	51.14	1987	17.03	1968	2.43
2005	47.21	1986	14.97	1967	2.25
2004	43.62	1985	12.64	1966	2.11
2003	38.73	1984	12.02	1965	1.95
2002	34.48	1983	11.59	1964	1.79
2001	33.20	1982	11.22	1963	1.63
2000	33.39	1981	11.34	1962	1.51
1999	32.36	1980	11.07	1961	1.41
1998	31.20	1979	9.82	1960	1.35
1997	31.32	1978	8.44		

The United States' GDP in 2008 and 2009 was $14.7 trillion and $14.4 trillion, respectively. For Canada, these numbers were $1.5 trillion and $1.4 trillion for the same period. The 2008 U.S. financial meltdown negatively affected not only its closest neighbor but also most developed countries and developing countries, as can be seen in the above table.

Table V.2 U.S. and Canada GDP in trillions of dollars
Sources: wwwradingeconomics.com/united-states/gdp/;
www.tradingeconomics.com/canada/gdp/

	2008	2009	2010	2011	2012	2013
United States	14.7	14.4	14.9	15.5	16.1	16.8
Canada	1.5	1.4	1.6	1.8	1.	1.8

I am reviewing the GDP of the past decades and earlier because of GDP's perceived association with the well-being of a society. For example, from Table V.2 one can see that the U.S. economy grew by $2.1 trillion from 2008 to 2013. This is not a small number; to put it in perspective, this was the GDP in Italy in 2013 or the equivalent to the GDP of California, the eighth largest world economy, in 2013. It is reasonable to assume that U.S. residents were better off, economically speaking, in 2013 than in 2008. In fact the economy has recovered from the Great Recession and employment has been on the rise, as witnessed by consumers' spending in goods and services. The sales of electronic products, facilitated by online trading, have substantially increased in the last ten years. This occurrence is one of the manifestations of a continuing need in all countries around the world for products, including electronic components. The U.S. sales of semiconductors, the foundation of computers and electronic products, reached $55 billion in 2011, mostly driven by the demand for cellular phones, tablets, and computers. It is generally accepted that in a free-market economy, incremental innovations intensify spending, which is reflected in GDP. In that sense, GDP provides a picture of the betterment of society, partial but acceptable as one of the facets of economic progress. One therefore can conclude that there has been some progress.

But the assumption that the higher a country's GDP, the better off its inhabitants has many limitations. There are many other factors not included in the GDP that could measure the level of satisfaction of people living in a particular country. Among its other inadequacies for assessing the well-being of a society, environmental costs are not taken into consideration. A recent article by Stewart Wallis of the World Economic Forum, entitled "Five Measures of Growth That Are Better Than the GDP" (10), identified: good jobs, well-being, environment, fairness, and health. Taken together, this framework sheds light on the degree of happiness and well-being of society in spite of the socioeconomic complexity and the subjectivity of these issues.

Following the initiative of Bhutan, the concept of Gross National Happiness (GNH) has caught momentum. March 20 has been declared International Day of Happiness (UN Resolution 66/281). Great Britain, France, and other countries have embarked on the concept to measure the degree of happiness of their citizens. Bhutan, a relatively small country of almost one million people in South Asia, considers GNH as important as GDP and has inspired many Western leaders to take a balanced approach to measuring the quality of life of their citizens.

In the United Nations' 2013 Gross National Happiness Report, the United States, although having the highest GDP in the world, did not occupy the first position. Denmark, with a GDP of 336 billion USD in 2013, ranked number one out of 52 countries for the most favorable living conditions. This assessment was based on a scientific survey covering a wide range of criteria. From a global perspective, Denmark's achievement can be seen as a sign of continuous economic progress.

Three other studies have been carried out: the OECD's Better Life Index (BLI), the Happy Planet Index (HPI), and the Social Progress Index. In the HPI, developed countries, including the United States, Canada, and European Union members, score high on life expectancy but low on ecological footprints. The Social Progress Index (2) has eliminated the traditional limitations of the GDP in the measurement of social progress by excluding economic variables and using outcome measures rather than inputs. In other words, how better off are people irrespective of how many resources have been used? Combining three dimensions—basic human needs, foundations of well-being, and opportunities—the report defines social progress as the capacity of a society to meet the basic needs of its citizens, establish the building blocks that allow citizens and communities to enhance and sustain the quality of their lives, and create the conditions for all individuals to reach their full potential.

Social progress tends to be a fluid concept subject to many interpretations, but not in the Social Progress Index because of the robustness of the component-level framework. A total of 133 countries have been studied, covering 94% of the world's population. This report provides one of the most comprehensive and accurate measure of social progress. The framework reflects, inter alia, a high consideration of inclusiveness.

In the Social Progress Index of 2013, updated in 2015, Norway ranked 1, Canada 6, and the United States 16. However, the United States ranked 8 in the opportunity category (see Appendix II).

Economic performance alone does not necessarily lead to a high ranking. The performance in other areas, such as nutrition and basic medical care, ecosystem and sustainability, tolerance and inclusion, is equally important. Nordic countries, among others having similar policies and performance, occupy the top 10 list of very high social progress achievement (Table V.3b)

Globally, on a population weighted basis, the average Social Progress Index scores 61.00, which is equivalent to the Social Progress Index of Kyrgyzstan, a country in Central Asia. It will be interesting to observe the movement of this average in future years. This could give us a global indication of a change in direction of society globally.

Table V3a Social Progress Index component-level framework

BASIC HUMAN NEEDS	FOUNDATION OF WELL-BEING	OPPORTUNITY
Nutrition & Basic Medical Care	Access to Basic Knowledge	Personal Rights
Water and Sanitation	Access to Information & Communication	Personal Freedom & Choice
Shelter	Health & Wellness	Tolerance & Inclusion
Personal Safety	Ecosystem & Sustainability	Access to Advanced Education

Table V.3b Social Progress Index 2015 results
Source: Social Progress Index. Reproduced with the permission of
socialprogressimperative.org

VERY HIGH SOCIAL PROGRESS INDEX		
RANK	COUNTRY SCORE	GDP PER CAPITA (PPP)*
1) Norway	88.36	$ 62,448
2) Sweden	88.06	$ 43,741
3) Switzerland	87.97	$ 54,697
4) Iceland	87.62	$ 41,250
5) New Zealand	87.08	$ 32,808
6) Canada	86.89	$ 41,894
7) Finland	86.75	$ 38,846
8) Denmark	86.63	$ 41,991
9) Netherland	86.50	$ 45,945
10) Australia	86.42	$ 42,831

*PPP: Purchasing power parity, which is an adjustment made to a currency to allow
comparability with other countries.

If we consider economic progress as the possibilities for citizens to sustain their quality of life and to reach their full potential, in this context out of 133 countries very few would pass this test. Many developed countries are not on the above list. Some people may agree that they are economically better off in 2015 than in 2005. After all, most countries have recuperated from the financial meltdown caused by the "subprime loans" fiasco, as well as high rates of unemployment. But it takes more to increase the quality of life of people so that they are capable of reaching their full potential. It is undeniable that there has been some economic progress but only benefitting the few, not the many. As mentioned earlier, the average social index of the 133 countries stands at 61. A real sign of progress will be the consistent year-over-year growth of this number for a certain number of years.

Change, as per Bertrand Russell's quote referred to at the beginning of this chapter, is an evolutionary process: the passage from one phase or stage to another. Progress is the development towards an improved and more advanced condition. If we associate progress with modernism, meaning that the world may be made better by human effort with the use of scientific knowledge and technology, perhaps this assertion will be acceptable to more people, including scientists, philosophers, and even some theologians. We can then look at progress as a human development process for the sustenance of material well-being via scientific knowledge and technological advancement. Since humankind is inseparable from technology, the degree of relationship between moral progress and technological progress will also be considered.

Technology as the expression of our determination to live goes back as long as we have existed. It is our will to thrive, and we have been moving forward for millennia. The case can easily be made to substantiate progress in technology. The statistics are staggering: in the health sector, an average person during the time of the Roman Empire might have expected to live only 25 years. By comparison, in 1900, the lifespan was 31 years and below 50 even in the richest countries. In 2013, the lifespan is over 80 years in most developed countries. Most infectious diseases, such as leprosy, smallpox, and polio, have been eliminated (9). More recently, finding a cure for the treatment of AIDS is on a good path. With the Brain Mapping Project currently underway, a cure will hopefully be found for the treatment of Alzheimer's and other brain diseases. If not, in the years ahead, because of the aging of the baby boomers, this disease may become dominant in our society. Already 44 million people worldwide suffer from Alzheimer's and other forms of dementia.

Communication has been enhanced around the globe. During the expansion of the Roman Empire, when soldiers left their families, they never saw each other again. Today audiovisual communication is taken for granted at the planetary and even extra-planetary level. The internet revolution has penetrated almost every country of our planet, making it possible to stay in touch with our friends and loved ones at a decreasing cost as time goes by. The same could be said for the transportation sector making continental and intercontinental travel convenient and affordable. One does not need to be a technophile to agree that substantial technological progress has been made. We have almost conquered all natural limitations. This includes extending (healthy) longevity, with the plausibility of conquering death in the future. Space conquest is a goal and will continue to be a work in process in the foreseeable future. We are even planning to live on other planets. It is evident that the universe's secrets are no match for the increasing power of human intelligence.

The agricultural revolution, the industrial revolution, and the digital revolution will remain important steps in the history of our society. Looking at those achievements, our ancestors would have considered them magical. Indeed, the famous futurist Arthur C. Clark, who anticipated the rapid expansion of technology, stated that any sufficiently advanced technology is indistinguishable from magic. Technological progress as witnessed by the advancement of knowledge and the improvement of the human condition is a reality, not a fiction or an illusion.

But it should also be noted that technological progress at the current stage is intimately linked to the fate of humankind. Progress in biotechnology, informatics, robotics, nanotechnology, and artificial intelligence suggests that wants and needs are being merged into a new concept, conveying "avid expectations." It seems that we are heading towards a capability to create whatever we want. But might does not always makes right. As mentioned by the futurist Ray Kurzweil in his book *The Singularity Is Near*, technology empowers both our creative and our destructive nature. Nanotechnology and artificial intelligence contain the seeds of our extinction as a species compared to the partial destruction of the atomic bomb. Even before the creation of gun powder, we used technology to destroy ourselves. Nanotechnology will take the medical field to a new level. Artificial intelligence is modernizing the manufacturing process, the transportation industry. The perception that the consequences of this increased efficiency and effectiveness may become uncontrollable was discussed in Chapter III.

In the analysis of technological progress I have also noted instances where inventions are used for a different purpose than planned by their creators, resulting in involuntary, unintended consequences. A common example is the occurrence of major unforeseen side effects of a prescribed drug in spite of it being properly administered. Another example is the voluntary withdrawal of a product based on complex field assessment results that show a negative impact on the environment. Because of such cases, the existence of oversight organizations and regulators is well justified. But as we all know, perfection is not within the reach of human beings. The case of DDT (dichlorodiphenyltrichloroethane), an effective insecticide used in World War II in the fight against malaria and thereafter, is troubling. Almost half of the world's population is at risk of this disease, which caused 429,000 deaths in 2015. This disease has been a scourge of humanity for millennia (3). An historical review of malaria conducted by Dr. Sophia Colantonio (Public Heath, University of Ottawa Canada) in her paper on the history of malaria indicates that Julius Caesar, Alexander the Great, the Spanish Conquistadors, and in the not too distant past, the construction workers of the Ottawa River Canal were affected by malaria. From other sources, it appears that part of Napoleon Bonaparte's overseas army was also decimated by this disease. There are approximately 200 types of malaria protozoa, but only four of them affect humans: plasmodium vivax, plasmodium falciparum, plasmodium malariae, and plasmodium ovale. A mosquito vector from the *Anopheles* genus is required to transmit the protozoan disease to humans through a bite to the skin. After the introduction of DDT, millions of lives were saved in Africa and tropical countries. Then came some ill-documented publications, including *The Silent Spring* of Rachel Carson, listing, inter alia, the negative impact of pesticides on bird populations. I say ill-documented because each of the arguments put forward has been refuted (6). DDT has been banned since 1972 not totally on a scientific basis but because this controversy became a political issue (7). It would have been more reasonable to ban the use of DDT in the Agricultural Sector. It appears that what is at stake is not technology itself, but the use of it if not guided by practical wisdom. DDT was not intended to be used as a pesticide. The belief that science does not always work in the best interest of society is sometimes due to the ramifications of misguided political interferences.

> "The human race has reached a turning point. Man has opened the secrets of nature and mastered new powers. If he uses them wisely, he can reach new heights of civilization. If he uses them foolishly, they may destroy him. Man must create the moral and legal framework for the world which will insure that his new powers are used for good and not for evil."

These were the words of the late U.S. President Harry S. Truman at the State of the Union Address on January 4, 1950. These words still resonate well today in view of the current development of emerging technologies. I am hopeful wisdom will remain our protective shield, the gist of our moral compass. This includes, among other things, virtue, justice, restraint, and empathy for others. It would be ideal if moral progress evolved in tandem with technological progress, but as observed by Isaac Asimov, science gathers knowledge faster than society gathers wisdom.

At the beginning of this chapter, the quote from Bertrand Russell indicates progress is said to be ethical. Basically, ethics is a set of moral principles that establishes what is good for an individual or a society. In that sense, ethics are important for the good functioning of a society. Since it was established earlier that morality has to some extent progressed in our society, the question then arises as to whether ethics in technology have also progressed. Ethics can sometimes raise divisive issues, particularly in view of ramifications in cultural, political, and economic spheres of activity and also because it is deeply rooted in the exercise of morality. The *Cambridge Dictionary* suggests that ethics and morality can be used interchangeably. Ethics deals with what is right, just, and fair, what ought to be given in a situation, an approach for making decisions outside of religion, the set of rules of a society. For an example, stealing is wrong, killing is wrong, coveting our neighbor's wife or husband is wrong, lying is wrong, loving our brothers and sisters like ourselves is right; those rules are written down and taught to us at an early age in our society. These are the guiding principles for our behavior.

When it comes to ethics, decision making sometimes can be complex and painful at the individual level because of the decision's uniqueness, the painful choice to be made. For instance, is it right to terminate the life of a suffering patient when current medical care is powerless? In other words, is it right for the family to observe, powerlessly, a loved one suffering for an indeterminate period? Is it right to kill one person to save many others? Is it right to accelerate the development of stem cells for curative purposes? Should the development of genetically modified organisms (GMOs) be extended to counteract famine and the effects of climate change?

The case of stem cell research is an interesting one because the current challenges that researchers face seem endless. Human embryonic stem cells can alleviate the suffering of patients as a result of an injury or disease because of the stem cells' self-renewal process and ability to differentiate and develop into all types of cells in the body. But here comes the ethical issue: the process for obtaining the stem cells by removing the inner cell mass of the blastocyst will cause the embryo to die. A multitude of issues are then opened for debate. For instance, when does human life start, since stem cells are collected three to five days after fertilization? Can a three-to-five-day-old fetus be considered human? Does the life saved outweigh the destruction of an embryo? An in-depth analysis of these ethical issues can be seen in the note (13) in the suggested readings section at the end of this chapter. I appreciate the position of the opponents of the use of embryonic stem cells even from a humanitarian perspective. Work is underway towards developing pluripotent cells from somatic cells with the same capacity as embryonic stem cells in the very near future. In other words, this ethical issue is temporary. In the meantime, it is my view that the removal of pain and suffering resulting from diseases or injuries in balance outweighs the current objections. As can be seen, making a decision from an ethical perspective can be difficult.

Another ethical issue concerns the use of the contraceptive pill for women. Approved by the FDA in 1960, acceptance was delayed by many setbacks from the beginning (in spite of the pill's benefits e.g., avoiding unwanted pregnancy, continuous careers for women, etc.). For a period of time in some developed countries, it could only be purchased by married women. From a purely ethical viewpoint, looking at the benefits of taking the contraceptive pill besides avoiding an unwanted pregnancy, three socioeconomic benefits can be identified: continuity in the labor force for women; counteraction of early marriage effects on education; and, from a global perspective, protection of resources, particularly in highly populated countries. The ethics bar has been lifted, and now billions of women around the globe are using the pill; rare are those institutions, besides the Vatican, raising a red flag about it. The position of the Pope, however, as the leader of the Catholic Church, can be understood from a spiritual standpoint. A leader does not need to be necessarily in agreement with the people. It does not matter to the Pope that 97% of sexually active Catholic women above the age of 18 have used some form of contraception banned by the Vatican (14).

Earlier, I used a quote stating that progress is ethical. In any assessment of progress, the past cannot be invoked to justify the present, as no one, for example, would want to go back to ancient, intolerant, labor-intensive, and insalubrious times. The belief in progress is about the future, a better future ethically and materially. As morality is a work in progress, technology is also a work in progress, subject to incremental changes obviously accelerating since the 20th century. In the history of society, there are issues so divisive that it seems impossible to reach any reasonable grounds for agreement; progress is one of them. The next decades will be very challenging from an ethical perspective. In the interest of freedom, justice, and human dignity, these issues will need to be addressed. Immobilism, or taking a fence-sitting position, will not make the issues go away. Among those are eugenics, cryonics, stem cells, cloning, brain implants, and voluntary human body modifications; but the more these issues are raised, the better our situation, because this signals the importance of addressing them if we want to maintain cohesion in a pluralist society. I suggest that ethics in technology has progressed but is temporarily overpowered by certain issues facing us now.

On October 18, 2014 Pope Francis declared that the Big Bang theory and evolutionary theory are real. God is "not a magician with a magic wand." For certain people, this declaration may facilitate a better appreciation of scientific realities; for others, this declaration will also enhance the development of new paradigms about strategic choices. I will close this chapter with a self-explanatory quote from Thomas Jefferson, and a word of wisdom from Catherine Pulsifer, author of many books, about the complexity of life and motivation. My position in all of this is to move forward with practical wisdom as our shield.

> "Some men look at constitutions with sanctimonious reverence, and deem them like the Ark of the Covenant, too sacred to be touched. They ascribe to the men of the preceding age a wisdom more than human, and suppose what they did to be beyond amendment. I knew that age well; I belonged to it, and labored with it. It deserved well of its country. It was very like the present, but without the experience of the present; and forty years of experience in government is worth a century of book-reading; and this they would say themselves, were they to rise from the dead. I am certainly not an advocate for frequent and untried changes in laws and constitutions. I think moderate imperfections had better be borne with; because, when once known, we accommodate ourselves to them, and find practical means of correcting their ill effects. But I know also, that laws and institutions must go hand in hand with the progress of the human mind. As that becomes more developed, more enlightened, as new discoveries are made, new truths disclosed, and manners and

opinions change with the change of circumstances, institutions must advance also, and keep pace with the times. We might as well require a man to wear still the coat which fitted him when a boy, as civilized society to remain ever under the regimen of their barbarous ancestors."

LIFE

Our entire life is made up of choices,
what we decide,
the action we take,
the attitude we display.
All represent the steps of life.

Sometimes we take two steps forward
and one step back.
Some of us take baby steps
some of us take giant steps

But the secret is not to let that
one step back turn into a failure.
Learn from backward steps

And keep on stepping forward in this dance
Called Life!

Suggested Readings

1. "How Canada Escaped the Global Recession," *Mises Institute*
   https://mises.org/library/how-canada-escaped-global-recession

2. "Social Progress Index 2015, Executive Summary," *Deloitte*
   https://www2.deloitte.com/content/dam/Deloitte/global/
   Documents/About-Deloitte/gx-cr-social-progress-index-
   executive-summary-2015.pdf

3. James Gips, "Towards The Ethical Robots"
   http://www.cs.bc.edu/~gips/EthicalRobot.pdf

4. The Precautionary Principle
   http://unesdoc.unesco.org/images/0013/001395/139578e.pdf

5. History of Malaria
   http://www.med.uottawa.ca/historyofmedicine/hetenyi/assets/
   documents/Sophia-Colantonio-Bloodletting.pdf

6. The Lies of Rachel Carson
   http://www.21stcenturysciencetech.com/articles/summ02/
   Carson.html

7. Bring Back DDT
   http://www.21stcenturysciencetech.com/articles/summ02/DDT.
   html

8. World Happiness Report 2012
   http://worldhappiness.report/ed/2012/

9. Thomson Prentice, "Health, History and Hard Choices: Global
   Health Histories"
   http://www.who.int/global_health_histories/seminars/
   presentation07.pdf

10. Five Measures of Happiness that Are Better than the GDP
    https://www.weforum.org/agenda/2016/04/five-measures-of-
    growth-that-are-better-than-gdp/

11. "List of Countries by Social Progress Index," *Wikipedia*
    https://en.wikipedia.org/wiki/List_of_countries_by_Social_
    Progress_Index

12. Eupraxsophy
    http://eupraxsophy.tumblr.com/abouteupraxsophy

13. "Ethics of Stem Cell Research," *Stanford Encyclopedia of
    Philosophy*
    http://plato.stanford.edu/archives/spr2013/entries/stem-cells/

14. "Catholics Stand Up and Speak Out for Contraception and
    Freedom of Conscience in Amicus Brief for *Zubik v. Burwell*,"
    Catholics for Choice
    http://cath4choice.org/news/pr/2016/
    CatholicsSupportContraceptionandConscienceinZubikBrief.
    asp

CHAPTER VI

A New Perspective

"The task is...not so much to see what no one has yet seen; but to think what nobody has yet thought, about that which everybody sees."
—Erwin Schrödinger

Every day that goes by, we are bombarded by news, including negative news about the risks of emerging technologies and the increasing gap between the haves and the have-nots. While some of those concerns are legitimate, they must be counterbalanced by the life improvements achieved in the last three centuries. That said, I am not totally praising the current economy, considering the prevailing income and wealth inequalities in most developed countries, particularly in the last 40 years. Inequality has a tendency to lead to political polarization in which a political party leader is perceived or claims, rightly or not, to be a populist, while at the other extreme a moderate right-wing leader proposes to improve the status quo, leaving the center confined to an ambiguous position. Only civic education can help people make a choice on their own in this challenging situation. A quote from another context, by the late Russian physician Anton Chekhov, can be used to explain the rise of populism: "Love, friendship, respect do not unite people as much as a common hatred for something." This could be applied to politically ambitious people. Currently, this "something" can be considered a euphemism for unemployment, the high cost of college or university, indebtedness, loss of home ownership etc. Too many virtuous citizens feel that they have been left behind by their governments, and they want a rapid change in their situation. Not only does a populist leader preaching along this line gain a receptive audience, but the rhetoric resonates so loudly that it adds fuel to a latent fire. For the sake of maintaining some cohesion in our society, inequality, this enormous politico-economic and self-reinforcing problem, needs to be addressed sooner rather than later. In the meantime, as an optimist, I am always hopeful because of the resilience and determination of humankind to pursue a higher quality of life than in earlier eras.

As mentioned in the preceding chapter, we are living, to some extent, in a better world: less violence, less poverty, longer life expectancy, better health, and, last but not the least, better technology. Our generosity and empathy for others are consistently shown when facing challenges such as pandemic diseases, environmental threats, and natural catastrophes. But there is an opinion in parts of society that the economy has not totally delivered on its promises of wealth distribution, which is a daunting task in itself. A few decades ago, a society of abundance seemed on the horizon, raising high expectations for all. What followed was a different world, fueled by the internet, mobile phones, and automation. One billion people still do not have access to the internet. While mobile phones are accessible in most parts of the world, their use is limited due to insufficient broadband in many countries. Automation has left many workers unemployed. Globalization has moved the point of production of goods and services for cost-efficiency purposes. This transformation has revealed to consumers and producers, from a transactional perspective, the smallness of the business world and thereby the reachability of everything. However, this rationalization has created a dislocation in the labor force, where the demand for highly skilled employees and the expansion of the service industry have become predominant.

A few years ago, the Venus Project Society (1), designed by the American futurist Jacque Fresco, created a lot of excitement. Having as a platform a resource-based economy (RBE), this approach was to promote a new alternative to the current economic system. Basically, profound changes would have to take place in many areas to facilitate its implementation, particularly in culture, technology, and property ownership. The proponents of this approach claimed that the combined effects of overconsumption, greed, fear, and lack of money are the sources of the current social inequities and woes. Therefore, the RBE's proponents suggest a complete reform by eliminating the use of fossil fuel and the commanding position of money at the forefront of our current socioeconomic system, for the sake of creating sustainable development throughout our planet.

A virtual tour provided on the Venus site makes this vision very appealing in the sense that nothing seems to be left out in the proposed new city model (e.g., energy, transportation, and housing). I must say that the project presents very persuasively a completely new direction and an ecologically sound future for society. To a certain extent, the inclusiveness element of this project and the different way of living could be seen as a direction towards social progress. However, the RBE in two of its core elements—the elimination of money and property ownership—is so radical that the whole concept may be difficult to implement.

In the same vein, a few years ago there was an alleged conversation about the creation of a single currency between the United States, Canada, and Mexico. Called the Amero, this theoretical currency would be similar to the Euro and would later expand to Central and South America. A single currency could greatly address the border issue, a sore point for the USA, and expand the free flow of goods, capital, and labor. The combined North and South America population market could reach close to one billion, compared to approximately 500 million for the 28 countries that use the Euro. One can immediately imagine the complexity of implementing such a currency system, including patriotism, the national debts of the potentially participating countries, the winners and the losers; the list could go on indefinitely. Patriotism is laudable, but empathy for other countries is also noble. We see manifestations of the latter in cases of natural catastrophes or in the assistance provided to restore democracy in many countries around the world. Numerous problems the world is facing are international. A fragmented approach is insufficient to solve them.

The fact remains that most countries are going through socioeconomic challenges and understandably may be predisposed to new solutions. The 2008 recession, technological changes, and competitive forces have affected all countries in the world to different degrees. In short, there is a general expectation for something better, a modification or even a reform of the current economic system to provide social and economic stability.

I am not a proponent of a total reform of the current free-market system for the sake of creating a new society. Those leaders who have tried to do so, particularly in the last century, have spectacularly failed and caused the abominable loss of millions of human lives. Implementing incremental economic policy changes can positively alter the current socioeconomic situation until the problems are resolved. There is no need for a theory of justice to decide what to do first. In summary, this is the position of Amartya Sen in *The Idea of Justice*. Although challenged by philosophers and economic theorists, in my view this approach provides a reasonable path to socioeconomic improvement.

I have come to accept the fact that social inequalities are not created by property ownership itself but by maladaptive socioeconomic policies and lack of competition, sustained by pre-established rules. Updating certain socioeconomic policies to cope with the source of the dissatisfaction is more practical and viable. In my view, Adam Smith's statement in *The Wealth of Nations* that "the reciprocal relationships that people voluntarily establish, channel self-interest to mutual advantage and promote a prosperous social order" still has some merits. Inherent in this statement, incorporating a wide spectrum of players, are (a) the individual freedom of a person to pursue his/her goals to the fullest and the freedom to choose the means to fulfill his/her needs; (b) the government's role in ensuring respect for contractual arrangements, being the guarantor of peace, order, national protection, and being the main provider of public goods. In the preceding chapters, I have argued that, after all, we have not done so badly considering the many challenges that society has faced in the last two decades. As a liberal egalitarian, what implementable policy modifications would I propose to improve life for all in the free-market system? In the following list, I have excluded education, as this item was addressed at the end of Chapter II. Rather than detailing these modifications, I summarize their core elements as follows:

1. Acceleration of Worldwide Connectivity

2. Readiness for Pandemic Diseases

3. Reduction of Poverty

4. Enhancement of Human Rights

5. Restoration of the Environment

I have approached this challenge from the perspective of the most pressing needs to be satisfied in the current reality. Furthermore, I have observed that those needs have a universal character in that they exist locally and globally. Each of the required policy modifications may be formulated differently, depending on whether the field of application is an undeveloped, developing, or developed country. I do realize that such an undertaking might sound illusory or like a wish list, in that the implementation of these five points is largely political. I suggest that this undertaking will survive the test of time, as globalization will continue to affect the society of each country. Any proposal of this nature will always remain ambitious because of political priorities or ideological disagreements. Successful implementation relies upon the governing powers of the day and support from international public institutions.

1) Acceleration of Connectivity Worldwide

As information technology continues to develop, connectivity will be a work in process affecting both developed and developing countries. At the developed-country level, there are major differences in price in fixed broadband plans, and the situation is the same for mobile broadband, with the resulting effect that according to the Open Technology Institute report entitled *The Cost of Connectivity 2014* (2), the majority of U.S. cities surveyed lag behind their international peers, paying more money for slower internet access. Although a number of factors, such as infrastructure and taxation, can explain the price differential between, say, the USA and Europe, it remains that service value for money, including in the suburbs, remains in need of improvement.

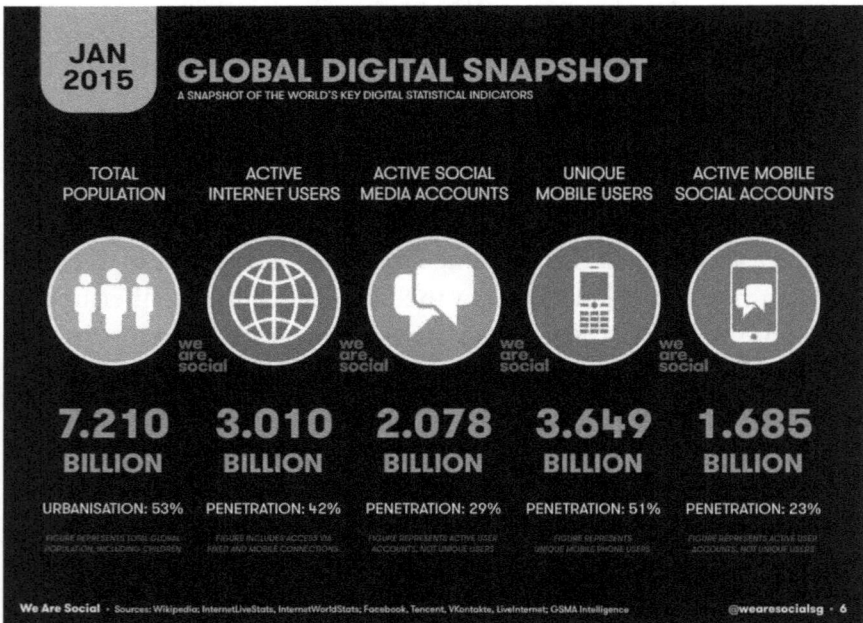

Figure VI.1a Communication modes Source: Wikipedia International

A substantial number of people around the globe are cell phone users. As shown in Figure VI.1a, the global market penetration of this device reached 51% in 2014. Active users of the internet is much lower, at 42.4% as can be seen in Figure VI.1b. The possible explanations are the fact that not everyone can afford a smartphone and/or there are broadband limitations. Much needs to be done for Central America and the Caribbean countries to increase their accessibility to the internet (Figure VI.1c).

As can be seen from these tables, a great discrepancy exists between the developed and the developing countries in the use of this unique instrument for communication, health, education, business opportunities, and thereby economic development. Connectivity facilitates the diffusion and implementation of new innovations.

The decreasing price of smartphones and the creative marketing strategies of internet services providers explain the global growth of the internet market. But more remains to be done not only to increase broadband but also to make the system sustainable in terms of the required infrastructure and power supply in some cases.

Table VI.1b World internet users in 2015

WORLD INTERNET USAGE AND POPULATION STATISTICS						
DEC. 31, 2014 - Mid-Year Update						
World Regions	Population (2015 Est.)	Internet Users Dec. 31, 2000	Internet Users Latest Data	Penetration (% Population)	Growth 2000-2015	Users % of Table
Africa	1,158,353,014	4,514,400	318,633,889	27.5 %	6,958.2 %	10.3 %
Asia	4,032,654,624	114,304,000	1,405,121,036	34.8 %	1,129.3 %	45.6 %
Europe	827,566,464	105,096,093	582,441,059	70.4 %	454.2 %	18.9 %
Middle East	236,137,235	3,284,800	113,609,510	48.1 %	3,358.6 %	3.7 %
North America	357,172,209	108,096,800	310,322,257	86.9 %	187.1 %	10.1 %
Latin America / Caribbean	615,583,127	18,068,919	322,422,164	52.4 %	1,684.4 %	10.5 %
Oceania / Australia	37,157,120	7,620,480	26,789,942	72.1 %	251.6 %	0.9 %
WORLD TOTAL	7,264,623,793	360,985,492	3,079,339,857	42.4 %	753.0 %	100.0 %

NOTES: (1) Internet usage and world population statistics are preliminary for Dec. 31, 2014. (2) Click on each world region name for detailed regional usage information. (3) Demographic (population) numbers are based on data from the US Census Bureau and local census agencies. (4) Internet usage information comes from data published by Nielsen Online, by the International Telecommunications Union, by GfK, local ICT regulators and other reliable sources. (5) For definitions, disclaimers, navigation help and methodology, please refer to the Site Surfing Guide. (6) Information in this site may be cited, giving the due credit to www.internetworldstats.com. Copyright © 2001 - 2015, Miniwatts Marketing Group. All rights reserved worldwide.
Source: www.internetworldstats.com

Table VI.1c Internet users in the Americas

INTERNET USERS AND 2014 POPULATION STATS FOR THE AMERICAS						
REGIONS	Population (2014 Est.)	% Pop America	Internet Users 30-Jun-2014	% Population (Penetration)	% Users America	Facebook 31-Dec-2012
North America	353,860,227	36.6 %	310,322,257	87.7 %	49.2 %	182,403,640
South America	406,194,811	42.0 %	230,727,557	56.8 %	36.6 %	142,708,440
Central America	164,210,961	17.0 %	72,373,646	44.1 %	11.5 %	48,933,540
The Caribbean	41,873,409	4.3 %	17,211,350	41.1 %	2.7 %	6,397,080
TOTAL THE AMERICAS	966,139,408	100.0 %	630,634,819	65.3 %	100.0 %	380,442,700

NOTES: (1) Internet usage and population statistics for the Americas were updated for June 30, 2014. (2) The Facebook subscribers were updated for December 31, 2012. (3) Click on each region to see detailed data for the individual regions. (4) Population numbers are based mainly on data contained in the US Census Bureau. (5) Internet usage stats come mainly from data published by Nielsen Online , ITU, Facebook and other local sources. (5) For methodology, definitions and navigation help, see the site surfing guide. (6) Data on this site may be cited, giving the due credit and establishing a link back to Internet World Stats . Copyright © 2014, Miniwatts Marketing Group. All rights reserved worldwide
Source: www.internetworldstats.com

The International Telecommunication Union (ITU) is the inter-governmental organization within which the public and private sectors cooperate for the development of telecommunications worldwide. One of the ITU's missions is to promote the development, efficient operation, usefulness, and general availability of tele-communication facilities and services. In the ITU report *The State of Broadband*, broadband is defined as 1.5 to 2.0 Mbps (megabytes per second), a bare minimum for a high-speed internet connection. However, in this age of information technology it is not uncommon to encounter in certain countries 256 kilobits per second (or 0.256 Mbps), a speed more akin to dial-up than to what much of the world would consider high-speed connectivity. Conceivably, competition between internet service providers will raise the bar. This is a necessity because slow speed combined with the limited purchasing power of subscribers in developing countries are impediments to them accessing world knowledge and business opportunities.

2) Readiness for Pandemic Diseases

Pandemic diseases, or the fear of them, have attracted a lot of attention lately. There does not seem to be a consensus-based single definition of pandemic diseases, except that they are infectious diseases that spread rapidly across many nations. Any discussion about pandemic diseases will suggest examples such as AIDS (acquired immune deficiency syndrome), SARS (severe acute respiratory syndrome) or, more recently, Ebola, originating in West Africa. As of October 22, 2014, the World Health Organization statistics have shown that 8,033 people were suspected to be infected with Ebola in West Africa, with 3,866 deaths.

But as unfortunate and deadly as Ebola was, the containment of this infectious disease was relatively effective, although Ebola is still being fought. More recently, the spread of the Zika virus in Brazil took the world by surprise. It appears that the global health system needs to be revamped to facilitate the containment of viral diseases (3).

In the same vein, potential human-made pandemic diseases should also be taken into consideration. This includes biological weapons, which involve the deliberate spreading of disease amongst humans, and "white plague," which is a pandemic disease developed by a frustrated scientist who vows revenge for some reason. Biological warfare could wipe out a substantial portion of the human population. Please, do not panic! For these particular problems without borders, there is a protocol in place to counteract them, particularly in developed countries. I have brought up readiness for pandemic disease as one of the five tenets to improve humankind's future because it will remain a continuing fight. The real test for readiness can be formulated by answering the following questions:

- Could information sharing between health partners about infectious diseases be improved?
- Is there sufficient laboratory capacity in pharmaceutical companies around the globe to deliver the appropriate vaccines and ant-viral drugs in a timely fashion?
- How effective is the standard communication protocol to timely inform each sovereign state around the globe about a potential pandemic disease?

■ At the individual level, how ready are we to face an outbreak of a pandemic disease?

I wish that I could provide a direct answer to each of these questions, because the life of millions if not billions of people could potentially be at stake. My observation is that there is a system in place to address the complexity of pandemic disease, including a protocol for the required measures to be taken, but their enforcement could be problematic and lengthy because of legal, ethical, political, and, to some extent, economic constraints.

3) Reduction of Poverty

The concept of poverty is relative to one's position geographically or economically. As such, there is a difference between relative poverty and absolute poverty from one country to another. Purchasing power below the national poverty line tends to be greater in developed countries.

Absolute poverty is defined as living on the edge of subsistence, with restrained access to basic human needs such as food, drinking water, shelter, education, and information. At a different level, relative poverty encompasses the issues of social exclusion and inequality on the basis of earned income. It is the inability to maintain an accepted standard of living in a particular country.

Poverty has accompanied humankind everywhere in the world for centuries if not millennia and is recognized by governmental institutions and most religious denominations. A comparative analysis of the extent of poverty between developed countries is difficult because of the lack of a universally accepted definition of the concept. This in turn brings limitations in the use of some international poverty lines. Canada does not have an official poverty line (4). However, a measure of absolute poverty in 2014, meaning the ability to obtain the basic necessities of life (food, clothing, shelter…) was calculated by the Fraser Institute* to be CAD 26,619 (USD 24,021) for a household size of four persons. Many other countries, such as Singapore and South Korea, do not support the poverty-line concept because of possible stigmatization and a perceived negative classification of their citizens. In the USA, the equivalent amount calculated by the U.S. Census Bureau (5) is $ 24,008. These numbers by no means infer a comparison. In fact, they are used for different purposes and converted to substantiate any particular situation. Nevertheless, the fact remains that poverty exists in those developed countries.

The World Bank has reported that there has been a substantial decline in global poverty (6). Almost one billion people still live in extreme poverty ($1.90 per day). In developing countries, the percentage was reduced from 52.2% in 1981 to 21% in 2010, with a further reduction to 12.7% in 2012. At the time of this writing, the projection for 2015 was 9.6%, or approximately 700 million people. Conceivably, extreme poverty can be eradicated in the next decade. However, for this descending trend to continue, more emphasis needs to be put on resilient growth, meaning growth deeply rooted in sustainable food supplies, energy, education, and specific local initiatives. I submit that it is crucial to eliminate extreme poverty from the viewpoint of human dignity at this stage of our civilization.

The fact that the extreme poverty elimination date has been postponed many times is not in indication of the failure of international aid organizations or the inefficiency of volunteers working strenuously around the world, but reflects the complexity of the notion of poverty when the political, economic, and social components are factored in. In this respect, the well-known economist Amartya Sen, in his book *Poverty and Famines*, explains that the lack of a means of food exchange is one of the most significant causes of poverty and famine. It also remains that famine is linked to political and ecological factors.

Aristotle once said that it is not easy for men to rise when their qualities are thwarted by poverty. Put in a different way, poverty is a great enemy to human beings, more specifically to human happiness. It makes the practice of some virtues difficult and other benevolent actions impossible. As Saint Thomas Aquinas wrote 700 years ago, "a minimum of comfort is necessary in life for the efficient practice of virtue." This is my rationale for suggesting the reduction of poverty including the elimination of extreme poverty, as one of the tenets for a better society.

But let's digress a bit.

Imagine for a moment that we suddenly receive a visit from an extraterrestrial being. After the usual ceremonial introduction at the United Nations, the nominated representative of our species comes to describe the fabric of our society, our socioeconomic structure, and what keeps us together. Let your imagination expand to a scenario in which you are watching this special and unique televised conversation between the alien and the representative of our species. From the intelligence gathered by the alien's sophisticated satellite, it then seeks clarification about the following persistent facts:

- More than 600 million people live without adequate shelter.
- Almost one billion entered the 21st century unable to read or sign their name.
- Almost one billion do not have access to clean water.

Although there is less poverty in the world at this writing than a decade ago, I have difficulty imagining the explanation that the nominated representative would provide, considering the fortune of the few wealthiest people in the world. Sixty-two people have the same amassed wealth as 50% of the poorest people on the globe. Thus far, we believe that we are alone in the universe, so this conversation remains fictitious. However, it has the merit of portraying an awful embarrassment. Poverty, and for that matter inequality, must be addressed with a concerted effort in order to solve this problem affecting our species.

In my search for the reduction of poverty, I have found a road less travelled that is worth exploring. The work of the Peruvian economist Hernando de Soto has been very inspiring and applies to the peculiarity of all developing countries. In his famous book *The Mystery of Capital: Why Capitalism Triumphs in the West and Fails Everywhere Else*, the author clearly and elegantly goes to the source of the current inequality between developed and developing countries, and his proposed solution is practical and doable. The combination of factors leading to the current deplorable situation of the less fortunate is primarily due to an abandonment of the building blocks of fortune. This mostly ignored and invisible rule of the free market even predates the invisible hand of Adam Smith. Basically, the acquisition and leverage of property is universally the source of wealth. It will take more than a few paragraphs to present the economic arguments of de Soto. However, for the sake of conciseness, I will only describe the requirements to achieve the end result, which is the reduction of poverty.

Having lived in many developing countries, and seen the applicability of his theory, I am convinced of its veracity. The crucial issue is the starting point. According to the World Bank, there are 500 million economically active poor individuals in the world excluded from the credit system. Acquisition and leverage of property for financing income-producing businesses could then make a big dent in the world poverty. In this context, property encompasses real property (land, buildings, and their improvement), personal property (belonging to one person), private property (belonging to a group of persons: business entities, natural persons), and intellectual property (exclusive rights over artistic creations, inventions, and copyright). Depending on the nature of the property, an owner has, among other rights, the right to consume, mortgage, sell, and transfer. The main document allowing the owner to do so is the title or right of ownership. This is the first challenge impoverished people in developing countries regularly face: the absence of ownership documents through neglect or technical limitations in the registry office. With the right ownership document, the owner can use the property to raise capital to finance his/her business. De Soto, in his above-mentioned book, has also corrected the fallacy that the inhabitants of less developed countries do not have property. They do, but they are confined in what de Soto calls dead capital, meaning not usable for raising capital because of the lack of a legal title. The inhabitants of those countries lack the process for making their property visible and thereby creating capital. For example, from de Soto's field assessments in Egypt and Haiti, he posits that the amount of dead capital is, respectively, $241.2 billion and $5.2 billion in those nations, enough to make the capitalist system work for them if it were recovered. But it is never too late to engage in modernization. After all, the integrated property system, as known in developed countries, is relatively recent. It takes time to be put such a system in place. In the 18th century, squatting has caused a lot of problems in the United States. De Soto has given the example of Japan, in which the widespread property system was completed only some 65 years ago in the early developmental period of that nation. This should give some hope to the inhabitants of less-developed countries.

4) Enhancement of Human Rights

All countries have laws to protect their citizens. These laws cover all aspects of human life and include social, political, and economic matters. Work, justice, education, and freedom resonate well in society members because they are inherent principles deeply entrenched in ourselves and indispensable for enjoying life. For instance, because of the concept of freedom incorporated in most nations' constitutions and in the UN's Universal Declaration of Human Rights (18), people can exercise freedom of speech, privacy, and many other rights. The vast number of signatories to the declaration reflects the degree of advancement of our current society, and the terms of this agreement also serve as the bar that international watchdog organizations use to determine cases of human rights violations.

Article 12 of the Universal Declaration of Human Rights states:

No one shall be subjected to arbitrary interference with his privacy, family, home or correspondence, or to attacks upon his honor and reputation. Everyone has the right to the protection of the law against such interference or attack.

Article 19 states:

Everyone has the right to freedom of opinion and expression; this right includes freedom to hold opinions without interference and to seek, receive and impart information and ideas through any media and regardless of frontiers.

I cite these articles particularly in the context of information technology (IT) and the challenges facing democratic countries to protect the civil liberties and privacy of their citizens. A substantial amount of work has been done thus far by concerned citizens' groups, business entities, and governmental institutions to address the privacy issue, particularly in the current digital age. Conceivably, IT sustains democracy. IT has made it difficult for a dictatorial form of government to maintain its grip on people forever. True is the saying that one cannot fool everyone all the time. We all can remember the Arab Spring and the subsequent conflicts, even when the rest of the world was cut off from the local news. IT facilitated the diffusion of information in real time during those challenging political events and contributed to people's ability to raise their human rights concerns at the international level. Although human rights are inalienable, they can be limited by the appropriate authority of a government when the action of some curtails the freedom of many, such as by crime and by incitement to violence and hatred.

Today, at the individual level, the invasion of personal privacy is a growing concern. A day doesn't go by without hearing that hackers or others have invaded private cyberspace for personal gain. Generally speaking, the minute that you log in, you are being watched (7). This is not to say that all tracking systems are bad. The case can be made for tracking systems that facilitate subsequent research on a particular subject.

I believe it is in the best interest of each individual to be aware of the sensitivity of the information that he/she uses in cyberspace. The privacy issue gets complex and undesirable when it comes to surveillance for national security purposes. I suspect that this opinion will vary in degree from one person to the next, depending on the individual's circumstance and the peculiarity of the situation.

I believe that no one would disagree for an instant with enhanced surveillance to prevent, for example, a repetition of the bombing at the Boston Marathon. The public's welfare is at stake. Additional potential situations are other acts of terrorism, child pornography, voluntary diffusion of pandemic disease, etc. In the hierarchy of rights, it is right to nudge the general public to accept the fact that some surveillance programs, such as email interception and wiretapping of conversations, are needed at the state level for the public's protection. Having said that, I am not entirely sure about how technically this need can be met, given the challenges raised by encryption, the number of service providers, and the multitude of communication software programs. The concept of good corporate citizen participation comes into play, since close cooperation between the corporate entities involved and the government is needed. I also believe that the chance of success of a surveillance endeavor would greatly increase if it were not overwhelmingly imposed. Cooperation instead of coercion could be a better path to follow.

A couple of years ago, I came across a paper entitled "Libertarian Paternalism Is Not an Oxymoron" (8), by Richard H. Thaler, a behavioral economist, and Cass R. Sunstein, a legal theorist. I believe that the approach taken in this paper could be applicable to information privacy. Libertarian paternalism is built on the foundation that a lack of information, or misinformation, or the structure of messages causes people to make choices that would otherwise be different. The authors posit that that there may be a way to help people make decisions in their best interest while at the same time serving the greater good. On the one hand, the libertarian element comes into play when an individual has the freedom of choosing between a few options, one of them being in his/her best interest. However, the architect of the set of options must allow the individual to opt out of the default option at minimal cost. On the other hand, the paternalism comes by nudging people in the direction of the public good. All other things being equal, it is the right of a public institution, in this case the government, to do so, resulting from the right to govern acquired through the election process. The less mandatory it is to nudge the individual to choose, the less paternalistic the initiative will be. I sense that most people would choose to protect their information. In that sense, there is a good "selling" job to do, considering past leaked information.

We all are very sensitive about privacy in general—for example, the privacy of correspondence, as in letters and emails, even from people we have not met. Henry L. Stimson, the U.S. Secretary of State in 1929, said in his memoirs, "Gentlemen do not read each other's mail." All of us have a wall around what we feel is very personal, things that we cannot share with others, regardless of the closeness of the relationship. We all feel invaded when someone opens our letters.

It is evident that there is a dilemma between the obligation of a nation to protect its citizens, its territorial integrity, and its economic structure against adversarial forces while also respecting the privacy conferred upon each individual by human rights legislation. I suggest that the cooperation of the population to protect their security should be sought. A population is a set of communities bound together with similar civic and sociocultural values. A long time ago, Sigmund Freud argued that shared identification and a sense of community are a better bastion of order than force. An analogy can be made with current community surveillance programs in which residents have mechanisms to obtain the protection of local authority against repeated strange intruders in a neighborhood. However, this cautious surveillance should not lead to the creation of a civilian police state in which each neighbor is watching the coming and going of each other's visitors.

The last challenge facing society in keeping up with human rights protection is the issue of morphological freedom (MF). This refers to the proposed civil right of a person to either maintain or modify his/her own body through informed, consensual recourse to, or refusal of, available therapeutic or enabling medical technology. There is quite a variety of opinion on this matter. A convincing argument by the transhumanist and futurist Anders Sandberg is provided in his paper (17) referred to in the suggested readings section of this chapter. He made the case that happiness, among other things, cannot be reached without morphological freedom. He argued that since happiness is a fundamental right of all humans and is guaranteed in Article 12 of the Universal Declaration of Human Rights, it then follows that morphological freedom is also protected. MF should be permitted and accessible if there is no potential harm to others. In addition, this extension of personal freedom is within the intent of human rights legislation.

In the same vein, one may recall the early green-and-yellow hair coloration of the mid-1960s. Call it new fashion or new wave, today people tend to express their personality in different ways. Marginal at the beginning, this style has become trendy over time. It also boils down to the notion of tolerance in a pluralist society.

5) Restoration of the Environment

Since the creation of the EPA (Environmental Protection Agency), followed by UNEP (the United Nations Environment Program), environmental concerns have gained momentum. In the preceding chapters, I have extensively written about the benefits of technology leading to the current state of society. However, this narrative would be incomplete if the environmental impact of industrialization were not factored in. Continuous use of fossil fuel, if no remedial measures are taken, will create an unsustainable way of living on this planet. Expressed by Greenpeace in the early 1970s, the protection of the environment has since taken many forms, all of them conveying the urgency of protecting the planet. In late 1997, the Kyoto Accord in the fight against global warming set the goal of reducing greenhouse gas emissions by 2012. Although the protocol was signed by most developed countries, the implementation went through some difficulties. Canada withdrew from the accord in 2011. The United States and China were not covered by the accord. Just before this book went to publication, under the Paris Accord of the United Nations Framework Convention on Climate Change (UNFCCC), dealing with greenhouse gases emission mitigation, an agreement was reached on a framework to reduce global warming. The participants agreed to keep the increase in the global average temperature to well below 2°C above pre-industrial levels and to pursue efforts to limit the temperature increase to 1.5°C above pre-industrial levels. There is a recognition that this goal would significantly reduce the risks and impacts of climate change. Perhaps recent ecological disasters have convinced world leaders to pay more attention to the effect of this manmade threat. They came to realize that climate change is real and, from an energy source viewpoint, business cannot continue as usual.

There are many dangerous myths about the environment. Bill Kovarik, in an article entitled "Environmental Issues Are Part of History" (November 4, 2010), mentioned a few of them. I am elaborating only on the first myth in the following list, leaving the rest to the opinion of others.

- The myth that Rachel Carson's 1962 book *Silent Spring* started all the uproar.
- The myth that environmentalism is just a hysterical reaction to science and technology.
- The myth that environmentalism is a passing fad with no serious ideas to offer.
- The myth that environmentalism is a substitute for religion.

The environmental protection movement has a long history in Great Britain. In the United States during the 19th century, proponents included Henry David Thoreau, who in his book *Walden; or, Life in the Woods* (1854) outlined the benefits of a close relationship with nature. The efforts of Yosemite Valley cumulated in the creation of Yosemite National Park by Congress in 1890. The environmental movement continued to grow in the 20th century, but not much was achieved after World War II until the publication of *A Sand County Almanac: And Sketches Here and There* in 1949 by the scientist and ecologist Aldo Leopold. He believed in a land ethic of maintaining the "beauty, integrity, and health of natural systems." That book achieved a record sale of two million copies and was translated into 12 languages. Rachel Carson is an important figure of the environmental movement. In her book *Silent Spring* (1949), Carson pointed out that air pollution killed 700 people in 1872 in the English city of Widnes and another 4,000 in 1952, precipitating the passing of the Clean Air Act in England. Her book also achieved a record sale of over two million copies. That said, and as mentioned earlier, part of the *Silent Spring* book dealing with the consequences of DDT remains questionable.

I particularly like the allusion to a farm boy crying wolf, made by former vice president of the United States and Nobel Prize winner Albert Gore in *The Future: Six Drivers of Global Change*: "In the parable of the boy who cried wolf, warnings of danger that turned out to be false bred complacency to the point where a subsequent warning of a danger that was all too real was ignored." I believe that environmental issues are real and must be addressed. It is known that since the First Industrial Revolution, the carbon dioxide concentration in the atmosphere has been skyrocketing from 280 ppm in 1750 to 400 ppm in 2015, an increase of approximately 40%. Figure VI.2 shows this trend. Obviously, this cannot continue forever without harmful effects on human life and biodiversity. In the face of this reality, people can either adopt a scientific skeptical attitude, meaning a critical thinking approach to find the truth, or total denial. The former is the proper attitude towards any challenge, while the latter is generally built on partial information or misinformation, with deplorable consequences (16).

Fig. VI.2 Greenhouse gas changes since the mid second century

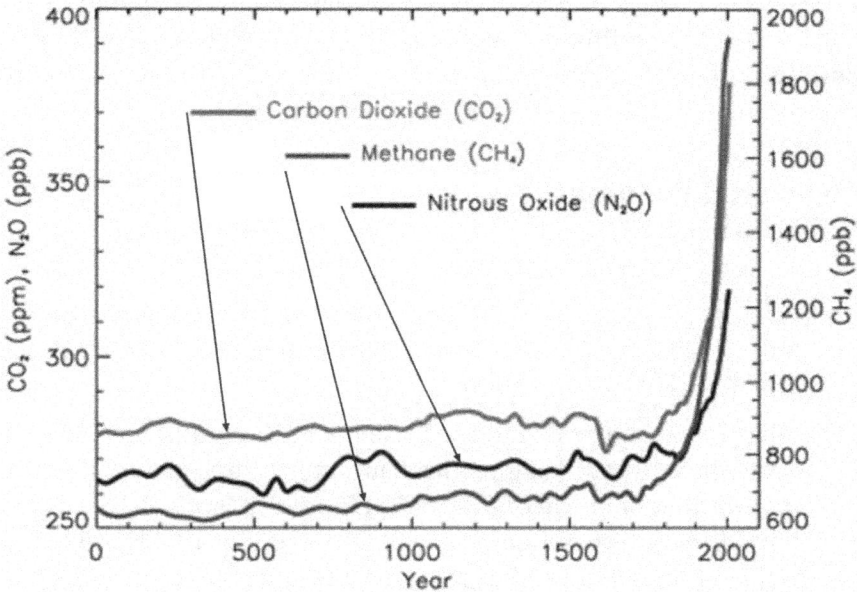

Source: American Chemical Society, https://www.acs.org/content/acs/en/climatescience/greenhousegases/industrialrevolution.html

On a positive note, even before the Paris Accord, some corrective actions have been taken. In developed and many developing countries recycling, increased use of public transportation, and renewable energy are widespread initiatives. New technological innovations are facilitating the management of toxic wastes that can be contained or disposed of at a lesser cost than a decade ago. I believe what is needed is an aggressive use of emerging technologies to address environmental issues. For example, it has been publicized since 2007 that nanotechnology can be used to reduce greenhouse gas emissions by the combined effect of many initiatives, such as nanoparticle additives in diesel fuel, solar cells, batteries, and super-capacitors for energy-efficiency purposes. So far, not that many applications in this particular field have been commercialized or the market penetration is moving slowly. Emerging technologies have been very visible in the medical and agricultural sectors. It is my hope that the environment will also soon be added to the list. Then we may enter a reduced carbon-emission era, and the technological growth sustaining renewable energy will result in new and durable economic development.

Suggested Readings

(1) The Venus Project
www.thevenusproject.com

(2) The Cost of Connectivity – Open Technology Institute
http://www.davidellis.ca/wp-content/uploads/2012/08/OTI_
The_Cost_of_Connectivity_2014.pdf

(3) "Zika Virus Outbreak and the Case for Building Effective and Sustainable Rapid Diagnostics Laboratory Capacity Globally," *International Journal of Infectious Diseases*, March 2016
https://www.researchgate.net/publication/297617108_Zika_
virus_outbreak_and_the_case_for_building_effective_and_
sustainable_rapid_diagnostics_laboratory_capacity_globally

(4) An Introduction to the State of Poverty in Canada
https://www.fraserinstitute.org/

(5) Institute for Research on Poverty, University of Wisconsin-Madison
http://www.irp.wisc.edu/faqs/faq1.htm

(6) Remarkable Decline in Poverty
http://www.worldbank.org/en/news/press-release/2015/10/04/
world-bank-forecasts-global-poverty-to-fall-below-10%-for-
first-time; major-hurdles-remain-in-goal-to-end-poverty-
by-2030

(7) Tracking Our Online Trackers
https://www.youtube.com/watch?v=f_f5wNw-2c0

(8) Libertarian Paternalism is not an Oxymoron
http://faculty.chicagobooth.edu/richard.thaler/research/pdf/
libpatlaw.pdf

(9) Anders Sandberg, "Morphological Freedom: Why We Not Just Want It but Need It"
http://www.aleph.se/Nada/Texts/MorphologicalFreedom.htm

(10) "Biological Warfare," *eMedicineHealth*
http://www.emedicinehealth.com/biological_warfare/article_em.htm

(11) "Pathogen," *Wikipedia*
https://en.wikipedia.org/wiki/Pathogen

(12) Pandemic Readiness Review
www.cidrap.umn.edu/news.../pandemic-readiness-review-says-4.5-billion-year-needed

(13) Institutional Readiness in Practice of Pandemic Response to an Emerging Infectious Disease
https://blogs.shu.edu/ghg/2014/06/16/institutional-readiness-in-practice-of-pandemic-response-to-an-emerging-infectious-disease/

(14) Strategies for Disease Containment– Ethical and Legal – NCBI
www.ncbi.nlm.nih.gov › NCBI › Literature › Bookshelf

(15) "The White Plague," *Wikipedia*
https://en.wikipedia.org/wiki/The_White_Plague

(16) The Scientific Guide to Global Warming Skepticism
https://www.skepticalscience.com/The-Scientific-Guide-to-Global-Warming-Skepticism.html

(17) Anders Sandberg, "Morphological Freedom: Why We Not Just Want It but Need It"
http://www.aleph.se/Nada/Texts/MorphologicalFreedom.htm

(18) Universal Declaration of Human Rights – United Nations
http://www.un.org/en/udhrbook/pdf/udhr_booklet_en_web.pdf

CHAPTER VII

Epilogue

"We know what we are, but know not what we may be."
—William Shakespeare, *Hamlet*

From 50,000 years ago to 2015, it has been estimated that approximately 107 billion humans have lived on this planet. As the world population stood at 7.2 billion in 2015 (1), we represent 6.8% of the total number of people who have ever lived on Earth (see Appendix I)

Figure VII.1 Human evolution

By chance or design, our species has survived many calamities: the Black Death, volcanoes, earthquakes, floods, and hurricanes. Our ancestors attributed natural calamities to the acts of gods or to the nefarious undertakings of bad spirits for the punishment of misbehaviors. This rationalization was their way of living with inexplicable events. To this day, we still fear the unknown or what we cannot understand, but the self-imposed guilt of ancient civilizations has been gradually displaced by the development of science and technology or, more specifically, by the diffusion of the idea of progress.

During our long journey on this planet, 99.9% of all the species that have existed are now extinct, some through our own doing, others because of natural calamities. We have beaten the odds by being part of the 0.1% remaining survivors, which quantitatively speaking are approximate 8.7 million (±1.3 million) species, 6.5 million on land and 2.2 million in the ocean's depths (2).

Our ancestors learned how to tame fire in 600,000 BC and invented the wheelbarrow in 400 AD. During this time span, human inventions continued to shape the world to come. Engineering, medicine, chemistry, mathematics, and astronomy were developing. At the beginning of the book, I associated humanity's fight to live with the determination of a stubborn boxer whose efforts are only encouraged by each knockdown. This fight is also our constant challenge with nature to eliminate the constraints and limitations preventing the betterment of our lives. Yet we are mindful that nature is also a great teacher, as we learn more and more about the growing processes of living organisms and the building blocks of matter.

From Archimedes, Copernicus, Galileo, and Newton to Faraday, Einstein, and William Shockley, the horizon keeps being extended by new paradigms. In the words of Robert Ornstein and James Burke (4), we have kept asking the axe maker (4) for better and more; and why not? That is our destiny. Technology, the expression of our consciousness, our volition to live as a species, materialized as a servant to us in the early period of our society, a friend supporting each industrial revolution and an ally, since we have matured enough to prevent ourselves, hopefully, from becoming extinct. I have underlined that each invention or innovation contains a constructive and destructive potential, depending on the user and the field of application, granted that human errors and their unintended consequences are inevitable. I have argued that wisdom should be one of our guiding principles to avoid falling in the trap of hasty and inappropriate choices.

Technology has permitted us to exit the confinement of our planet to explore the immensity of space outside of our galaxy and our place in the universe. It is still unknown whether other life forms exist or have existed outside of our solar system. The search for extraterrestrial intelligence (SETI) is ongoing. As a windfall of space exploration, we are now in a position to identify and destroy asteroids that are on a collision course with our planet. We can shatter them like throwing boiling water onto ice. Exploration is in our nature. We began as wanderers and we are still wanderers, said the famous futurist Carl Sagan. The continuing quest to go to the farthest point of the universe seems to be encoded in our genes. *Homo sapiens* expanded from East Africa to Asia, Europe, and America. Now, space is the next frontier.

In the preceding chapters, I have provided my observations and analysis of the materialization of our will to live and improve our lives, from the Renaissance to the first decade of this century, mindful of the substantial contribution of those who have proceeded along the same path, such as the ancient Sumerian and Chinese civilizations. My purpose for doing so was to facilitate an understanding of the evolutionary character of technology and the ephemeral nature of each developmental stage. From linear to exponential growth, this evolution reflects our preferences and expectations and sometimes even our irrational exuberance for better and more. Like a wheel in motion, the concept of inevitability that I invoked in Chapter II characterizes this bijective relationship between humans and technology. We affect it, and it has been affecting us consciously and unconsciously as long as we have existed. The First and Second Industrial Revolutions drew people from the suburbs to the cities, liberating us from hard, repetitive work and thereby creating new lifestyles. The third revolution, also called the digital revolution, holds the promise of transforming our economic life, social habits, and legal systems in ways of which we are unaware, just as as the Fourth Industrial Revolution used the digital revolution as a spring board.

The rewards and challenges of technology for society is the theme of this book. Indeed, it is clear that scientific progress has occurred and, as a result, society is better off today than two and a half centuries ago. However, we are facing various challenges. Some innovations appear to be divisive at the ethical level: stem cells, germline intervention, cloning, longevity, artificial intelligence, robotics, and personal privacy, to mention a few. Taken individually, each one has the capacity to substantially change society as we know it today. But here is the kicker. These innovations also contain the seeds for a different and more manageable future in which we will have more options to exercise. I remain confident that these dividing issues will be addressed by technology, because if the latter seems to be part of the problem, it will certainly be part of the solution. It becomes then of utmost importance to strive to be informed about technological development in general so that we can have input about the solution. It is my hope that the preceding chapters and the appendices will contribute to that end.

On the topic of being open-minded about new ventures and choices, the last time I was in the Newark (USA) airport, I took a driverless train to go from terminal A to terminal C. It also happened that I sat in the first compartment, overlooking the cabin of the non-existent driver. The drive was smooth, and the stop at each designated station was very precise and timely. Will I use a commercial pilotless airplane to fly from Toronto to Paris? Probably, after reading the flight tests results and scientists' opinions. I have debated these issues in Chapter V and concluded that we must strive to maintain an analytical approach as a quintessential element of our behavior as science and technology continue to develop.

But at the end of the day, the most important remaining question is: Where is humanity heading? I am not a philosopher, anthropologist, or, for that matter, even a scientist in any discipline. I am a generalist, a common wanderer of our species, moving forward along a trajectory that has suddenly reached a curve onto an unexpected path. This path, like our own history, which never flows in a predictable way, provides a panoramic view of a possible destination. In essence, this is our journey on this planet: a trajectory to a better place and, ultimately, a better world.

Throughout the writing of this book, each chapter has raised in me a range of emotions varying from humility to pride. Meditating on the first users of this trajectory, I was deeply touched by the creativity, respect, and love displayed by *Homo erectus* (5) in their sociality of eating together facilitated by the use of fire, their caring for the weak and disabled, and the burial of their dead, including with flowers. It remains unknown whether they believed in an afterlife. They also made more and better tools than their ancestors *Homo habilis*, such as wooden bowls and spears to survive and to improve their life during their long journey following the same trajectory. This type of caring is unique to human beings and has been instilled somehow from them to us for more than 400,000 years. It goes without saying that this type of behavior preceded the establishment of religion. A more elaborate form of behavior is being preached today by the major religions in our evolved and sophisticated society. Among the dogmas being promoted are respect for and love of others, altruism, eliminating suffering, and helping fellow humans to improve their lives on Earth, with the rewards of an afterlife, within their interpretation of this word. Belief leads to faith, then to confidence and trust. Whether religious belief is the exclusive impulse of human beings to create a better world or the influence of a super-spiritual force to achieve this goal does not really matter. Both momenta converge to the same final point. In a tolerant and pluralistic society, we witness the influence of both. This brings me to state that the practice of religion and the support of technology are not mutually exclusive.

Transhumanism, a movement supporting technological development for the betterment of humankind, as a concept goes back to the Renaissance, with Francis Bacon or even earlier, and was revisited in the 18th and 19th centuries by William Godwin but better articulated by August Comte and Nikolai Fyodorov. All three wrote at some point about a profound conviction regarding the dominion of humankind over nature and the improvement of society. The ongoing creative process, as time goes by, leads to a greater and higher good for all.

In an article in *The Journal of Evolution and Technology* (December 2008, Vol. 20, Issue 1), Eric Steinhart described the philosophy of the Jesuit priest Pierre Teilhard de Chardin (7), many aspects of which can be found in the transhumanism movement. In Teilhard's philosophy, the universe is evolving towards a godlike state, the highest level of perfection of humankind: the Omega Point, the ultimate state of perfection of humankind. This long and inevitable transformative process is not necessarily incompatible with any religious belief, because technology is one of the means toward this end point.

The futurist and transhumanist Ray Kurzweil alluded to the same thing in *The Singularity Is Near*:

> *Evolution moves towards greater complexity, greater elegance, greater knowledge, greater beauty, greater creativity, and a greater level of subtle attributes such as love. In every monotheistic tradition God is likewise described as all of these qualities, only without any limitations: infinite knowledge, infinite intelligence, infinite beauty, infinite creativity, infinite love and so on... The accelerating growth of evolution never achieves an infinite level, but as it explodes exponentially it certainly moves rapidly in that direction. So evolution moves inexorably towards this conception of God, although never quite reaching this ideal. We can regard, therefore, the freeing of our thinking from the*

severe limitations of its biological form to be an essentially spiritual undertaking.

Some transhumanists are atheists and some are not. However, if the focus is on the end point, I suggest that it becomes difficult to argue that it will not benefit humankind. In Chapter II, I mentioned religious belief as one of the factors at the source of indifferent or negative attitudes towards technology. When machines gain the ability to improve themselves independently through the development of artificial intelligence (AI), which in essence is the core of the singularity concept, certain scholars believe that anything could happen. The fear is that humankind may become the slave of those intelligent machines, confined to execute their decisions in all fields of human affairs. This is a possible outcome of emerging technologies if we don't cautiously manage their development and limit them to being friendly allies. From a dystopian perspective, if we still exist, we will regress rather than progress. It is encouraging to know that some safeguards still are in place to prevent the realization of this scenario.

If a reluctant attitude towards AI and nanotechnology is a matter of prudence, it infers a state of rationalization, understandingly a process preceding decision making in each developmental stage of these technologies, which is wise. If, however, it is a matter of conscience in the sense of right and wrong, I suggest that AI and nanotechnology, for that matter technology as a whole, will contribute to the reachability of the Omega Point mentioned earlier. Conceivably, our universal values are being embedded in the design of these intelligent machines in order to prevent an "out-of-control" situation. The betterment of life since the first axe maker has preceded any religion; technological development is religion-neutral or, rather, a nonreligious issue.

Supporters of technology adhere to the transhumanist philosophy. The latter can be viewed as an extension of humanism, from which it is partially derived. Humanists believe that society matters, that individuals matter. We might not be perfect, but we can make things better by embracing rational thinking, freedom, tolerance, and empathy for our fellow human beings. Transhumanists agree with this concept but also emphasize what we have the potential to become infinitely better. Just as we use rational means to improve the human condition and the external world, we can also use such means to improve ourselves, the human organism, not limited by traditional humanistic methods, such as education and cultural development. The use of technological means may eventually enable society to move beyond the limitations of what some would think of as human (8).

Faith is personal, and the exercise of it is guaranteed in the constitutions of most democratic countries. I will conclude this section with a practical quote from Dag Hammarskjold, the second Secretary of the United Nations: "I would rather live my life as though there is a God and die to find out that there isn't, than to live my life as though there is no God and die to find out there is."

As I mentioned in the Prologue, this book emerged from my reflections about technology, its socioeconomic ramifications, and its future direction. Whatever has happened or will happen is part of an inevitability in the pursuit of our destiny. Yes, we do have a destiny, as per the evolutionary path moving towards greater perfection, learning from each previous experience. At the beginning of this chapter, I quoted a passage from Shakespeare's *Hamlet*: "We know what we are, but know not what we may be." This long journey on planet Earth is of finite duration when the natural components sustaining human life are factored in. For instance, the sun has enough energy to shine for another five to six billion years. Needless to say, we don't know what our morphology will be that far in the future. From an evolutionary perspective, it would be like comparing ourselves to our pre-hominid ancestors. Surely by then, things will be very different, caused by the choices we make. This century is of crucial importance for humankind. In the meantime, assuming that an existential risk does not materialize, we will have developed ways and means to control the components of our environment and possibly to populate the universe.

Back to our present time, as the economy continues to develop, propelled by new ideas and technologies, it is my view that developed countries will experience an unprecedented era of well-being. This new state of affairs will provide a new template that will also benefit developing countries. But there is another windfall for developed countries, as most work will be transformed by automation. As time goes by, intelligent machines will become more and more a convenience or even a necessity in our daily lives, thus reducing our work and making recreational activities more possible and attractive. Moreover, as a result of the combined effect of increased productivity resulting from leadership and technology, we may realistically approach the dawn of a leisure society, coveted since 1932 by John Maynard Keynes: "three hours a day is quite enough to satisfy the old Adam in most of us." This will not be the end of work but the expansion of opportunities to discover ourselves and our place in the universe. Perhaps then, as an evolved species, we will come to more deeply appreciate the value of life and cooperation between all the nations on Earth.

Suggested Readings

(1) Current World Population
http://www.worldometers.info/world-population/

(2) Carl Haub, "How Many People Have Ever Lived on Earth?" *Population Reference Bureau*
http://www.prb.org/Publications/Articles/2002/
HowManyPeopleHaveEverLivedonEarth.aspx

(3) C. Mora, D. Tittensor, S. Adl, A. Simpson, and B. Worm, "How Many Species Are There on Earth and in the Ocean?" *PLOS Biology*
http://journals.plos.org/plosbiology/article?id=10.1371/
journal.pbio.1001127

(4) James Burke, Robert Ornstein, "A Report on *The Axemaker's Gift*"
www.humanjourney.us/axemaker.html

(5) Homo Erectus, Fire, Tools and Culture
factsanddetails.com/world/cat56/sub360/entry-2754.html

(6) "What Does it Mean to Be Human?" *The Human Journey*
http://www.humanjourney.us/

(7) Teilhard de Chardin, Pierre. *The Phenomenon of Man*. Harper Perennial Modern Classics, 2008

(8) Transhumanism FAQ
http://humanityplus.org/philosophy/transhumanist-faq/#answer_19

(9)) "Biological Warfare," *eMedicineHealth*
http://www.emedicinehealth.com/biological_warfare/article_em.htm

(10) "Pathogen," *Wikipedia*
https://en.wikipedia.org/wiki/Pathogen

Notes and Glossary

The notes and glossary are intended to explain some of the terms used in the book, and they include my own comments. They are not presented in alphabetical order.

Artificial Heart

The artificial heart stems from the need to provide patients who have incurably defective left and right ventricles a safe alternative while waiting for a heart transplant. There are two types of total artificial heart (TAH): one is connected to an outside power source and the other is not. The humanitarian factor plays an important role in the choice to use a TAH; the patient can spend time with his/her family while waiting for a transplant. So far, receivers of TAHs have survived close to 18 months and as long as five years. The TAH has been used more frequently in recent years because of the combined effects of a decreasing number of donors and the potential for rejection of the new organ by the recipient's immune system. Shutting down the immune system is not a viable option, as the patient may die from infection. A TAH is also an option for people who are not suitable for heart transplant.

Artificial Intelligence

Considering the abundant literature on this subject in the recent past, artificial intelligence (AI) will remain a hot topic for years to come, particularly because of its ramifications in our affairs. A machine capable of behaving like or better than a human incites as much emotional response as, for example, immortality or cloning. Amidst the myriad definitions and in an effort to shed light on the meaning of AI, let's look at each term separately.

Artificial means unnatural, synthetic, and refers to a replication, human creation, or innovation. For example, we all are aware of natural and artificial vanilla and the differences between the two; I say this without falling into the trap of the naturalistic fallacy that something is good or right because it is natural, or bad or wrong because it is unnatural. Shedding light on what artificial means is the easy part.

Defining (*human*) *intelligence* is more complicated because of the complexity of the biological and mental linkages. I submit that intelligence is the innate ability to know, recognize, remember, create, communicate, and respond to past, current, or future situations. One can see in this definition a tripartite activity composed of a learning process, a collation process, and a reasoning process, all governed by thought processes emanating from the various sections of the brain.

Putting both definitions together (artificial and intelligence), I suggest that AI is a human-made artifact or machine capable of performing independently any intellectual or physical task at or above the level that a human being can perform; creativity is also inferred.

Up to now, there have been three levels of AI: *narrow artificial intelligence* (NAI), also called weak artificial intelligence; *artificial general intelligence* (AGI), also called strong artificial intelligence; and *artificial superintelligence* (ASI).

NAI is already part of our daily life, such as in automatic translation programs and computer games. AGI is more sophisticated and not limited to simulating human tasks and behavior. AGI is now in its infancy. Examples of AGI are facial recognition and Google driverless cars. Many challenges exist in the wide range of emotional, sudden, unpredictable situations that AI machines may face and their computational or automated responses. The ability to subjectively feel an emotion, react to an inner perception, and display the appropriate behavior are the effects of neurological connections that are not yet embedded in AGI machines.

We are still far away from a sentient machine (agent). I submit that those cases may not be addressed until the BRAIN (Brain Research through Advancing Innovative Neurotechnologies) Initiative, in the USA, and the Human Brain Project, in Europe, are completed by 2030 and more computational power is achieved. One of the goals of the BRAIN Initiative is to understand the relationship between the structure and functions of the human brain (see www.brainmapping.org).

Artficial superintelligent machines are still far down the road. In his 2006 article "How Long Before Superintelligence?", the philosopher Nick Bostrom defined ASI as an intellect that is much smarter than the best human brains in practically every field, including scientific creativity, general wisdom, and social skills. This is the stage where machines will enormously exceed human mental capacity. Given the physiological limitations of the human brain compared to the expandable memory of a machine, various scenarios are foreseen, such as brain implants and brain uploading to take advantage of this new know-how. The promises of such a technology are so deep that I would say they qualify as another evolutionary phase of humankind, but there are perils if wisdom is not intertwined in the process. In this respect, it is encouraging that many writers have already forewarned us about the dangers and the potential unintended consequences of the availability of such an advanced intellectual capacity. AI is and will continue to be the subject of heated debate over the next decades. For a comprehensive overview of the subject, I also suggest reading "The AI Revolution: The Road to Superintelligence," Parts I and II, by Tim Urban, at waitbutwhy.com. This article is very informative and easy to read.

The Path to Complexity

In most literature about the development of the universe, reference is often made to the second law of thermodynamics because it contains the concept of entropy, coined by the German scientist Rudolph Clausius. The first law is the concept of energy conservation: energy cannot be destroyed or created. The second law of thermodynamics states that as energy is transferred or transformed, more and more of it is wasted. It also means that there is a natural tendency of any isolated system to generate into a more disordered state. Entropy can therefore be understood as a lack of order or predictability. The entropy of any system that is not in thermal equilibrium almost always increases.

Source: https://www.livescience.com/50941-second-law-thermodynamics.html

Fuel Cell

The physicist and chemist Sir William Robert Grove invented the first fuel cell in 1839. Wikipedia defines a fuel cell as:

"A device that converts the chemical energy from a fuel into electricity through a chemical reaction of positively charged hydrogen ions with oxygen or another oxidizing agent. Fuel cells are different from batteries in that they require a continuous source of fuel and oxygen or air to sustain the chemical reaction, whereas in a battery the chemicals present in the battery react with each other to generate an electromotive force. Fuel cells can produce electricity continuously for as long as these inputs are supplied…. Individual fuel cells produce relatively small electricity potential. So cells are 'stacked' or placed in series to create sufficient voltage to meet an application's requirements."

Fuel cells will play a major role in the manufacturing of electric cars and will reduce toxic gas emissions into the environment.

Cloning

Cloning is the process of producing biologically identical organisms by making copies of DNA fragments, cells, or other entities, such as plants, animals, or bacteria, leading respectively to three types of cloning: recombinant DNA cloning, therapeutic cloning, and reproductive cloning.

Recombinant DNA cloning is a process in genetic engineering of transferring to a host cell genetic material from one or multiple organisms, thereby creating sequences that would not otherwise be found in the original genome. One of the most common applications is the transfer of insulin from a patient to the DNA of a second organism. This organism then becomes an insulin production factory.

Therapeutic cloning is the use of stem cells to replace cells or organs damaged as a result of disease. These new cells or organs are grown from stem cells. Basically, the nucleus of an egg is extracted from the skin cell of a patient requiring stem cell treatment and inserted in the egg of a donor; the nucleus of the donor's egg has previously been removed. After chemical or electrical stimulation to divide, a cluster of cells is formed and the inner part of this cluster is isolated to create embryonic stem cells, which are infused into the patient for treatment. The process will become totally "clean," ethically speaking, when the donor's egg is replaced by a human cell to avoid destroying an embryo. This intervention does not reproduce a human being. In therapeutic cloning, the unused stem cells remain in the laboratory for research purposes.

Reproductive cloning, according to the Center for Genetics and Society, is the "production of a genetic duplicate of an existing organism." The process for producing an embryo in therapeutic and reproductive cloning is similar with one major exception. In reproductive cloning, the embryo is inserted in the uterus of the donor to achieve a genetic replicate of the animal from which the body cell was taken. The replica has the same characteristics as the host, although not always. Reproductive cloning is currently used in agriculture and for research, and perhaps will be used in the near future for the reproduction of endangered species. Most countries have specific laws forbidding the replication of human beings.

Cosmic Time Scale

What is the cosmic time scale? Imagine the entire history of the universe compressed into one year, with the Big Bang corresponding to the first second of New Year's Day and the present time to the last second of December 31 (midnight). Using this scale of time, each month would equal a little over a billion years. Carl Sagan has compressed into one year the time from the creation of the universe to the preceding century. On the basis of that scale (1) the first human appeared on December 31 at 10:30 pm. From this perspective, the appearance of *Homo sapiens* is relatively recent.

Source: (1) http://visav.phys.uvic.ca/~babul/AstroCourses/P303/BB-slide. htm
Source: (2) https://en.wikipedia.org/wiki/Cosmic_Calendar

Emerging Technologies

Emerging technologies cover the inter-application of a wide range of disciplines to achieve a desired result. It was found that combining the increased development of many disciplines could better address the scientific challenges of the past 50 years than a fragmented approach. Emerging technologies include, but are not limited to, educational technology, information technology, nanotechnology, biotechnology, cognitive science, robotics, and artificial intelligence. At first glance, one can see from the above list the contribution of the internet to the sciences of informatics, physics, chemistry, biology, physiology, and psychology. Various acronyms are sometimes used to identify this convergence of technologies in certain applications, for example: NIBC (nanotechnology, informatics, biotechnology, cognitive science), BRAIN (biotechnology, robotics, artificial intelligence, nanotechnology), and GRIN (genetics, robotics, informatics, nanotechnology). The future social impact of these converging technologies is enormous. As is always the case, the innovations resulting from these converging technologies carry their dual seeds of construction and destruction, one of the reasons there is so much emotion about the future.

Encyclopedia

The encyclopedia of Denis Diderot and Jean le Rond d'Alembert emerged as one of the finest products of the Enlightenment period, not only because of the laudable objective of making available to the world the knowledge of humankind but also because the structure and content of the document allowed people to learn and think freely. It was produced by the collaborative effort of 140 writers, who included philosophers, mathematicians, and other intellectuals. It remains difficult to determine the far-reaching consequences of this masterpiece for the entire European population. However, it is commonly agreed that the political and religious hegemonies were weakened because the encyclopedia of Diderot and d'Alembert stimulated free thinking and provided a new perspective on reality.

Fast forward to Wikipedia, the free online encyclopedia, which is written and edited by anonymous volunteers working without pay, with the objective of making available human knowledge to everyone. Due to the internet, according to the statistics, approximately 500 million people visit the Wikipedia site every month. The dream of Diderot is being realized because of progress in information technology.

Generations of Computers

The need for a simplified and rapid calculating mechanism has been around for thousands of years. From the invention of the abacus by the Babylonians between 1000 BC and 500 BC to Thomas de Colmar's calculating device called the arythmometer in 1845, all show convenient machine-computing capability that exceeds the speed of human intellectual calculations.

In 1832, the mathematician, philosopher, and mechanical engineer Charles Babbage invented the first mechanical computer and the concept of a programmable computer. It had a central processing unit (CPU) and 1.7 kilobytes of expandable memory, and the program could be made using punched cards. For many reasons, including finances and personality conflicts, this computer was not built in Babbage's lifetime. In 1991, a mechanical computer based on his design was built; subsequently, it was displayed at the London Science Museum in 2002, and at the Computer History Museum in Mountain View, California, in 2008. It weighs five tons, contains eight thousand parts, and measures 11 feet long by seven feet high.

Another important name in the development of computer programing is the English mathematician, logician, and computer scientist Alan M. Turing. By creating a machine capable of performing most computing tasks, Turing played an instrumental role in the development of algorithm concepts and computations. In 1950, he developed the Turing test, substantiating that a computer can simulate a natural process such as the human thinking process. In his landmark paper "Computing Machinery and Intelligence" (1950), he stated that the Turing test displays a machine's "ability to exhibit intelligent behavior equivalent to or indistinguishable from that of a human." He planted the seeds of artificial intelligence, which is still being developed.

Since 1950, computers have evolved, and in the computer industry they are grouped as generations. Vacuum tubes with the use of machine language software, transistors with the use of low-level programming language software, integrated circuits with the use of structured high-level programming language software, and micro-processors with the use of domain-specific high-level programming language software represent, respectively, the first, second, third, and fourth generations of the computer. The fifth generation includes computers capable of voice, image, and graphic recognition aided by artificial intelligence. They are able to respond to natural language and solve many problems at the same time. I consider the sixth generation to be in the making. With the development of more powerful computers, actual artificial superintelligence may be reached within the next decades. Since the fourth generation of computers, reduction in price and size and increased performance have been the hallmarks of all electronic devices, particularly computers and mobile phones.

How Many People Have Ever Lived on Earth?

This semi-scientific calculation was made by the Population Reference Bureau. It is obvious that prior to the 19th century, many estimations had to be made because of the lack of data and their reliability when available. The difficulty resides in the selection of the starting period for assessing human populations (here set at 50,000 BC), leading to different numbers for the Earth's total population over time. The methodology of the approach taken by the sources referenced hereunder is appropriate and makes it possible to provide answers to this common question of public interest.

Year	Population	Births per 1,000	Births Between Benchmarks
50,000 B.C.	2	-	-
8000 B.C.	5,000,000	80	1,137,789,769
1 A.D.	300,000,000	80	46,025,332,354
1200	450,000,000	60	26,591,343,000
1650	500,000,000	60	12,782,002,453
1750	795,000,000	50	3,171,931,513
1850	1,265,000,000	40	4,046,240,009
1900	1,656,000,000	40	2,900,237,856
1950	2,516,000,000	31-38	3,390,198,215
1995	5,760,000,000	31	5,427,305,000
2011	6,987,000,000	23	2,130,327,622

NUMBER WHO HAVE EVER BEEN BORN	107,602,707,791
World population in mid-2011	6,987,000,000
Percent of those ever born who are living in 2011	6.5

Source: Population Reference Bureau estimates.

http://www.prb.org/Publications/Articles/2002/HowManyPeopleHaveEverLivedonEarth.asp
http://www.worldometers.info/world-population/

Insulin and Vaccines

I have grouped these two inventions of the 20th century because of their enormous contributions to our well-being.

Insulin

No one can now imagine a world without insulin. Based on 2014 statistics, in North America alone 31 million people die every year of diabetes types I and II: 29 million in the United States and two million in Canada. In the United States, the cost of lost productivity due to diabetes was $69 billion in 2013. Invented by F.G. Banting and C.H. Best in 1921, insulin was first administered in 1922 to a dying boy in Toronto, Canada. Insulin allows diabetics to have a normal life in spite of the malfunctioning of their pancreas. Equally interesting are the current and future improvements in the delivery methods of insulin. From inhaled insulin to tablets, these delivery methods will eliminate the use of needles, which will likely have a positive impact on compliance.

Vaccines (Polio)

Polio (poliomyelitis) is a highly infectious viral disease that may attack the central nervous system and is characterized by symptoms ranging from a mild, non-paralytic infection to total paralysis in a matter of hours. It most commonly infects children and older adults. At its peak in the 1940s and 1950s, polio paralyzed and killed over half a million people worldwide every year, including tens of thousands in Canada and over 300,000 in the United States.

Vaccinations have changed our world, as witnessed by their success against diphtheria, smallpox, and polio. The polio vaccine is credited to Jonas Salk and was first used in 1952. The invention or discovery (the case is still being debated) of vaccines remains one of the most effective methods for mass protection against pandemic diseases.

The Internet and the World Wide Web

I thought it appropriate to define the internet and the World Wide Web together, as they are both related to connectivity yet remain different.

Internet

In my summary of the most important inventions of the 20th century (see Chapter I), I stated that the internet was invented by ARPANET in 1983. While this is true, the creativity of many institutions and people, particularly the American psychologist and computer scientist Joseph Carl Robnett Licklider, preceded the creation of the internet. His idea was to create a network wherein many different computer systems would be interconnected to one another to quickly exchange data, rather than have individual systems set up, each one connecting to some other individual system (http://www.todayifoundout.com/index.php/2014/09/history-internet/). This was in 1963.

The contributions of Vincent Cerf and Bob Kahn in 1974 must also be acknowledged for providing TCP/IP, the Transmission Control Protocol/Internet Protocol to facilitate the transmission and reception of data in proper order. In 1983, TCP/IP became the official communication protocol of ARPANET. Many authors consider this date to be the creation of the internet. Today, Wikipedia defines the internet as the global system of interconnected computer networks that use the internet protocol suite (TCP/IP) to link billions of devices worldwide.

Wikipedia in its article "Internet" (7) quoted Jonathan Strickland in *How Stuff Works* (June 27/2014) as saying that the internet has no centralized governance in either technological implementation or policies for access and usage; each constituent network sets its own policies. True. No one owns the internet, or rather, thousands of people and organizations own bits and pieces to control quality and access levels.

Eric Schmidt in *The New Digital Age: Reshaping the Future of People, Business and Nations* stated that the internet is among the few things that humanity has built that humanity doesn't understand, our largest experiment in anarchy. This statement should not be taken too literally but refers to the internet as a complex network system connecting computers all around the world, the ultimate instrument of humankind for free speech, not governed by a central authority or individual. Having said that, while this wonderful instrument expands democracy, the internet can also be used to incite assaults upon democracy.

World Wide Web

The World Wide Web is a system of internet servers that support specially formatted documents. The documents are formatted in a language called HTML (hypertext markup language) that supports links to other documents, as well as graphics and audio and video files. The web was invented by the English computer scientist Timothy John Berners-Lee in 1991. More specifically, this date refers to the time that the first website built was put online. Berners-Lee is credited with the creation of HTML and the hypertext transfer protocol (HTTP) for distributing information on the internet. Within a particular website, connections can be made to multiple other sites through the internet via hyperlinks.

The internet allows communication from computers to computers, and the World Wide Web (a part of the internet) allows multiple connections to other websites through the internet. The internet can function without the web (email is an example), but the web cannot operate without the internet.

Longevity

For thousands of years, the issue of life extension and, for that matter, immortality itself has been a never-failing source of interest for humanity. As evidenced in the oldest known literary work, the Epic of Gilgamesh, dated 2100 B.C., King Gilgamesh of Uruk (currently Iraq), overwhelmed by the death of his friend, sought from the gods the secret of eternal life. The active search for the fountain of youth over the many centuries reflects humans' constant preoccupation with achieving prolonged adulthood.

Fast forward to the new millennium. With advances in biotechnology and computer science, longevity has hurtled to the forefront of scientific debates. A thorough investigation is underway in various parts of the globe about the various causes of death, particularly aging, along with the related religious and ethical considerations. Indeed 150,000 people die every day, two-thirds due to aging. A quote from the biomedical gerontologist Aubrey de Grey encapsulates the current attitude towards aging in the research community: "Aging is a barbaric phenomenon that shouldn't be tolerated in a polite society." His work focuses on the prevention of the aging process through an approach summarized in the acronym SENS (Strategies for Engineered Negligible Senescence).

Many other promising approaches are also being pursued, but no solution has yet been found. The challenge resides in the complexity of the aging process and the multitude of correlations with the cell production mechanism in the human species after maturity. Peter Medawar (1952) has defined aging as the collection of changes that render human being progressively more likely to die. João Pedro de Magalhães on his website prefers the term senescence instead of aging and refers to the latter as a progressive deterioration of physiological function, an intrinsic age-related process of loss of viability and increase in vulnerability. Others, including the author of this book, claim that aging is a transformative process following maturity, leading ultimately to physical limitations and deteriorative appearance caused by neglect and inherited traits. In these definitions out of many, one can detect two elements: the accumulation of physiological damage as aging advances, and the consequences of genetic inheritance. In the first case, aging appears as unnatural, a disease that needs to be cured; this is often referred to as the damage-based theory. In the second case, an intrinsic aging process can be construed to mean an inborn or congenital condition. Any intervention to reverse this process could be seen as an attempt to alter the individual genetic program. Known as the programmed theory, one can see the potential ethical issues raised in the context of each of these approaches. This is debatable because, as documented by João Pedro de Magalhães in his paper "Programmatic Features of Aging Originating in Development: Aging Mechanisms beyond Molecular Damage?" (http://www.fasebj.org/content/26/12/4821.long), aging changes are also the result of regulated processes, some forms of antagonistic pleiotropy (when one gene controls more than one trait where at least one of these traits is beneficial to the organism's fitness and at least one is detrimental to the organism's fitness).

Both the damage-based and the programmed theories must be examined in the search for a cure to aging, taking into consideration the contributions and the limits of each these approaches. From a humanitarian perspective, it is my position that aging can and should be cured. I support the argument that naturalness cannot be invoked, as aging is not always evident in other species. Aging is not a normal evolutionary process. In fact, it is independent from evolution. The latter aims at the reproduction of the fittest, not the survival of the fittest.

The aging process is very complex, but sooner or later, bio-gerontologists will come up with a solution to this nuisance affecting humankind. In the meantime, there is a confusion, at least in my opinion, between people's perceptions of longevity and infinite lifespan (immortality). Longevity is a long, healthy, and active life, free of physical limitations and deterioration. Prolonged adulthood sounds better than an infinite lifespan because it carries a more positive connotation. Immortality would then be the outcome of a prolonged adulthood. In longevity and immortality, humankind will still not be immune to death, because of extrinsic factors, such as accidents (cars, explosions) or natural disasters (earthquakes, volcanoes, hurricanes). Immortality can also be achieved by other mechanisms, such as brain uploading (we are not there yet) or (reproductive) cloning. This latter option is forbidden in most countries. Conquering death will be one of the last frontiers of humankind.

Mechanical Clock

It is widely accepted that the transformation of textiles from a home-based industry to factory production and the refinement of the mechanical clock increased the productivity and profitability of the textile industry in Western Europe in the 17th century. The strict regulation of working hours and the supervision of workers paved the way to the First Industrial Revolution. Textile workers could no longer work at their will to meet their own financial needs; instead, they had to work according to the goals of the business owner. At this point, because of the use of the clock, time became a commodity that could be compressed, sliced, or expanded to meet a production level. Time itself had become feared or adulated, depending on one's economic position. As mentioned in Chapter III, the Canadian writer George Woodcock in the *Tyranny of the Clock* emphasized the negative impact of this invention as an infringement on individual liberty. Indeed, how time is felt is a matter of debate. Be that as it may, time measurement continues to be one of the key instruments used to improve productivity and profitability.

Movable Type Printing Press

It is said that the movable type printing press invented by Johannes Gutenberg in the 15th century ushered society into modernity. A movable type printing press had existed many centuries before Gutenberg, since the Song dynasty (960–1270) in China, the first government to issue paper money. This then raises a question about the originality of the Gutenberg printing press. Its originality resides mostly in its efficiency. Although block printing and movable metal type presses were known in Korea and China, the 50,000 characters of the Chinese language did not make the movable type printing press practical.

Gutenberg, besides his own additional component creations, is credited for streamlining and refining an efficient mass-production printing system. Basically, printing entails the application of pressure to an ink surface on a print medium (paper, for example) for the production of books, pamphlets, or other literary material for publication. While screw presses and paper were invented in the first century, the creativity of Gutenberg is revealed in the creation of a lead-based alloy for his pieces of type (letters, images). Moreover, he achieved mass production by inventing a special mold for casting the letters. He is also credited with the invention of an oil-based ink for high-quality metal printing because the water-based ink then used was soaking the paper.

The invention of the movable type printing press accelerated the mass production of literary material throughout Europe.

Newton's Universal Laws of Motion and Gravity
The Three Universal Laws of Motion

1. An object at rest will remain at rest unless acted on by an unbalanced force. An object in motion continues in motion with the same speed and in the same direction unless acted upon by an unbalanced force.

2. Acceleration is produced when a force acts on a mass. The greater the mass (of the object being accelerated), the greater the amount of force needed (to accelerate the object).

3. For every action there is an equal and opposite reaction.

The Universal Law of Gravitation

Any two bodies in the universe attract each other with a force that is directly proportional to the product of their masses and inversely proportional to the square of the distance between them. (Source: en.wikipedia.org)

Singularity

Although acceleration and complexity have made most concrete developments impossible to predict, large-scale statistical factors, such as wealth, life expectancy, intelligence, productivity, speed of information transmission and speed of information processing, increase in a surprisingly regular way. This makes it possible to extrapolate their development into the future. Most of these growth processes are exponential, characterized by a constant doubling period, or a constant increase in percentage per year. This means that the underlying growth mechanism is stable, producing a fixed number of new items for a given number of existing ones.

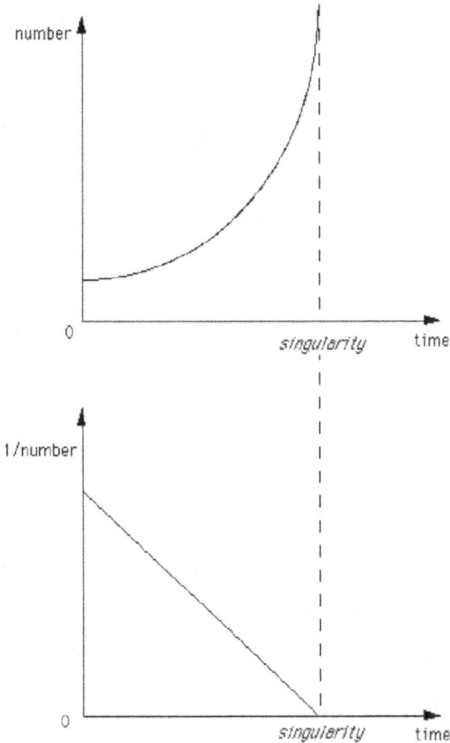

Fig. 1: A hyperbolic growth process, where the growing number becomes infinite in the singular point, while its inverse (1 divided by the number) becomes zero.

However, some processes grow even more quickly. For example, population growth in percentage per year is much larger now than it was a century ago. This is because medical progress has augmented the gap between the percentage of births and the percentage of deaths per year. If the growth of the world population over the past millennia is plotted on a graph, the pattern appears to be hyperbolic rather than exponential. Hyperbolic growth is characterized by the fact that the inverse of the increasing variable (e.g., one divided by the total population) evolves according to a straight line that slopes downward. When the line reaches zero, this means that the variable (world population, in this case) would become equal to one divided by zero, which means infinity. This is essentially different from an exponential growth process, which can never reach infinity in a finite time.

If the line describing the inverse of world population until recently is extended into the future, we see that it reaches zero around the year 2035. Of course, such an extrapolation is not realistic.

It is clear that world population can never become infinite. In fact, population growth is slowing down at the moment. However, that slowdown itself is a revolutionary event, which breaks a trend that has persisted over all of human history.

In mathematics, the point where the value of an otherwise finite and continuous function becomes infinite is called a "singularity." Since traditional mathematical operations, such as differentiation, integration, and extrapolation, are based on continuity, they cannot be applied to this singular point. If you ask a computer to calculate the value of one divided by zero, it will respond with an error message. A singularity can be defined as the point where mathematical modelling breaks down. The inside of a "black hole" is an example of a singularity in the geometry of space. No scientific theory can say anything about what happens beyond its boundary. We can only postulate that different laws will apply inside, but these laws remain forever out of sight. Similarly, the Big Bang, the beginning of the universe, is a singularity in time, and no amount of extrapolation can describe what happened before this singular event.

The mathematician and science-fiction writer Vernor Vinge in "The Coming Technological Singularity" has proposed that technological progress is racing towards a singularity; scientific discovery appears to be an exponentially growing process with a doubling period of about 15 years. However, the doubling period is diminishing because of increasingly efficient communication and processing of newly derived knowledge. The rate of growth is itself growing. This makes the process super-exponential, and possibly hyperbolic. Vinge would argue that at some point in the near future, the doubling period will reach zero, which means that an infinite amount of knowledge will be generated in a finite time. At that point, every extrapolation that we could make based on our present understanding would become meaningless. The world will have entered a new stage, where wholly different rules apply. Whatever remains of humanity as we know it will have changed beyond recognition.

Some data supporting this conjecture can be found in a chart provided by Peter Peeters, in which the time elapsed between a dozen fundamental discoveries and their practical application is plotted against the year in which the invention was applied. The graph shows a downward sloping line, which reaches zero in the year 2000. These are just a few data points, and the implication that inventions will be applied virtually instantaneously after 2000 does not yet mean that scientific progress will be infinite. Vinge himself would situate the date of the singularity between 2010 and 2040. His reasoning is based on the accelerating growth of computer-aided intelligence. Rather than considering the IQ of an isolated individual, he would look at the team formed by a person and computer. According to Vinge, a PhD armed with an advanced workstation should already be able to solve all IQ tests ever devised. Since computing power undergoes rapid exponential growth, we will soon reach the stage where the team (or perhaps even the computer on its own) achieves superhuman intelligence. Vinge defines this as the ability to create even greater intelligence than oneself. That is the point at which our understanding, which is based on the experience of our own intelligence, must break down.

A related reasoning was proposed by Jacques Vallée. Extrapolating from the phenomenal growth of computer networks and their power to transmit information, he noted that at some point, all existing information will become available instantaneously everywhere. This is the "information singularity."

These models should not be taken too literally. They are metaphors, proposed to stimulate reflection. It does not make much sense to debate whether "the singularity," if such a thing exists, will take place in 2029 or in 2035. Depending on the variable you consider most important, and the range over which you collect data points, you may find one date in which infinity is reached, or another, or none at all. Even if nothing as radical as a "New Age" can be predicted, the zero-point method is certainly useful for attracting attention to possible crisis points, where the dynamics of change itself are altered. Peter Peeters has used this method with some success to "predict" historical crises, such as the First World War and the 1930 and 1974 economic recessions. (The Second World War, strangely enough, did not seem to fit the model.)

The point to remember, however, is that abstract, non-material variables, such as intelligence, information, or innovation, aren't subject to the same "limits to growth" that characterize the exhaustion of finite resources. Such variables could conceivably reach values which, for all practical purposes, may be called "infinite." Several parallel trends show a hyperbolic type of acceleration that seems to reach its asymptote (the point of infinite speed) somewhere in the first half of the 21st century. This does not mean that actual infinity will be reached, only that a fundamental transition is likely to take place. This will start a wholly new mode of development, governed by laws that we cannot as yet guess.

References: P. Peeters, *Can We Avoid a Third World War in 2010?* Vernor Vinge, "The Coming Technological Singularity"

"The Socio-Technological Singularity," by Francis Heylighen, PhD
Reproduced with the permission of the author.

Steam Engine

In the history of technology there are many inventions for which the inventorship is disputed. The steam engine is one them. Its origin goes back to the first century; however, our focus is not on its origin but on the contribution this device has made to our society. The game-changer was that it made the extraction of coal possible by pumping steam below the groundwater surface. As Europe's main source of fuel became coal, the Thomas Savery steam engine modifications in 1698 allowed the extraction of coal at a deeper level, where water was more abundant. Then along came Thomas Newcomen, who created the first commercial steam engine in 1711, which was capable of producing steam to operate a water pump. James Watt's modifications by the end of the 18th century extended the use of the steam engine beyond mining. From agriculture to manufacturing and transportation, applications of the steam engine occupied a primary role in the First and Second Industrial Revolutions.

Stem Cells

Stem cells are unspecialized cells that can differentiate into specialized cells and divide to produce more cells. In mammals, there are two broad types of stem cells: embryonic stem cells and adult stem cells, the latter being called somatic cells. The former are found in the inner part of a structure (blastocyst) formed four to five days after fertilization or from the umbilical cord just after birth, while the latter can be taken from various tissues, including bone marrow and blood but excluding eggs and sperm. An induced pluripotent cell is a stem cell that has all the characteristics of an embryonic stem cell with the exception that it is a specialized adult stem cell taken from certain organs of the body and transformed through the introduction of genes that reprogram it. It is not known yet whether in their application, induced pluripotent stem cells are totally equivalent to embryonic stem cells.

For a complete glossary of stem cells and related components, please visit the website of the National Health Institute at: http://stemcells.nih.gov/glossary.htm

Technology

Throughout this book, I have defined technology as the concrete expression of humankind's volition to live. In my view, this expressivity has taken different forms with the passing of time. The visible effect was first seen in a servant relationship through the rudimentary work of instruments and animal-powered devices, followed by a friendly relationship in the mechanization period, which was followed by an ally relationship in the automation period continuing today. This last period is the apex of a relationship: two entities having a common goal for the well-being of humankind. It is my hope that it will continue that way.

The Three Laws of Robotics

In 1942, Isaac Asimov introduced the three laws of robotics in the science-fiction story "Runaround":

1. A robot may not injure a human being or, through inaction, allow a human being to come to harm.

2. A robot must obey the orders given it by human beings except where such orders would conflict with the First Law.

3. A robot must protect its own existence as long as such protection does not conflict with the First or Second Laws.

Isaac Asimov, *Handbook of Robotics*, 56th Edition, 2058 AD

Asimov introduced a fourth law in 1985 called the zeroth law (number 0), preceding the other three in robots' interaction with humanity: A robot may not harm humanity or, by inaction, allow humanity to come to harm.

Other rules have been added subsequently by others to prevent ambiguity in robots' decisions, disobedience, or unintended action/inaction. This last is one of the concerns about robots. They are perceived to be equipped with an intelligence level higher than and different from ours, a potential source of mistrust.

Transistor

It is my view that the transistor is the most important invention of the 20th century. For the sake of simplicity, I have retained the transistor definition from dictionary.com as an electronic device that controls the flow of an electric current, most often used as an amplifier or switch. Transistors usually consist of three layers of semiconductor material, in which the flow of electric current across the outer layer is regulated by the voltage or current applied at the middle layer. Invented by William Shockley, Walter Brattain, and John Barden in 1948, this tiny device has revolutionized the electronics industry by making possible a diminution of size and price together with a performance increase in all electronic equipment. Without the transistor, the building of integrated circuits and microprocessors would not be possible. This is the brain of laptops, communication devices, and other electronic products that we currently enjoy using. Millions of transistors are produced every year for this purpose. In that sense, it remains the most radical innovation of the 20th century. Moore's Law states that the number of transistors in a dense integrated circuit will double every two years. As of today, this prediction has proven to be accurate.

Choice

I have to conclude with choice, out of the alphabetical order of the glossary, because of the importance of this act that we constantly commit, voluntarily or not, in the course of human affairs. I want the concept of choice to be remembered because, driven by culture, inherited traits, environment, and psychological or physical needs, choice determines our future. Simply formulated, choice can be defined as the act of selecting a course of action between two or more options.

We sometimes adopt past practices because of the persistence of old paradigms. For example, the form of a bicycle seat is less comfortable than that of a motorcycle, from an ergonomic design standpoint. We still choose the former, perhaps because of its popularity or, better, the influence of the distant past saddle in the bicycle designer's mind. Moreover, have you ever wondered why the letters QWERTY are always in the topmost row of letters on your computer keyboard? One of the legends quoted by the late Stephen J. Gould of Harvard University is that originally this layout not only prevented the type bars from jamming but also allowed salespeople to quickly type the name of the first production line: typewriter. Since then, this layout has become the automatic standard choice for all keyboards, from computers to mobile phones. In this respect, the phenomenon of path dependency, largely mentioned in history and the social sciences, can be invoked. It portends that past events influence present and future actions. I bring these examples to raise awareness of the powerful forces that we encounter in our decision-making process, sometimes involuntarily. We are all the product of our beliefs, education, and past experiences. That said, it must also be noted again that technology, like history, never flows in a predictable way.

In the first half of this century, and surely long thereafter, lifestyle will be complex because of the quantity, quality, and variety of new products and services coming on the market. They will be very different from what exists now, and I mean *very* different. Although we cannot separate ourselves from our past, we should be open-minded and predisposed to assessment and rationalization to the greatest extent possible before embracing or rejecting a course of action. Another important point to remember is that while we are free to choose a course of action, we must also be prepared to live with its consequences at the individual and/or collective level.

May practical wisdom* be used in our choices to sustain the flourishing of humankind.

---

* Practical wisdom in Aristotelian philosophy is equivalent to *phronesis*. Specifically, this is a Greek word for a type of wisdom relevant to practical things (such as the right use of technologies), requiring the ability to discern how or why to act virtuously, and the ability to reflect upon and determine good ends consistent with the aim of living well overall.

Major Inventions of the 20th Century and the First Decade of the 21st Century

Inventions of the 20th Century

1900

- Count Ferdinand von Zeppelin invented the zeppelin.
- Charles Seeberger redesigned Jesse Reno's escalator and invented the modern escalator.

1901

- King Camp Gillette invented the double-edged safety razor.
- The first radio receiver successfully received a radio transmission.
- Hubert Booth invented a compact and modern vacuum cleaner.

1902

- Willis Carrier invented the air conditioner.
- James Mackenzie invented the lie detector or polygraph machine.
- The teddy bear was born.
- George Claude invented neon light.

1903

- Edward Binney and Harold Smith co-invented crayons.
- Michael J. Owens invented bottle-making machinery.
- The Wright brothers invented the first gas-motored and manned airplane.
- Mary Anderson invented windshield wipers.
- William Coolidge invented ductile tungsten, used in light bulbs.

1904

- Thomas Sullivan invented teabags.
- Benjamin Holt invented a tractor.
- John A Fleming invented a vacuum diode or Fleming valve.

1905

- Albert Einstein published the Theory of Relativity and made famous the equation, $E = mc^2$.

1906

- William Kellogg invented Corn Flakes®.
- Lewis Nixon invented the first sonar-like device.
- Lee Deforest invented the electronic amplifying tube (triode).

1907

- Leo Baekeland invented the first synthetic plastic, called Bakelite.

1908

- Elmer A. Sperry invented the gyrocompass.
- Jacques E. Brandenberger invented cellophane.
- J.W. Geiger and W. Müller invented the Geiger counter.
- Fritz Haber invented the Haber process for making artificial nitrates.

1909

- G. Washington invented instant coffee.

1910

- Thomas Edison demonstrated the first talking motion picture.
- Georges Claude displayed the first neon lamp to the public on December 11 in Paris.

1911

- Charles Franklin Kettering invented the first automobile electrical ignition system.

1912

- Motorized movie cameras were invented, replacing hand-cranked cameras.

1913

- Arthur Wynne invented the crossword puzzle.

- The Merck Chemical Company patented what is now known as ecstasy.
- Mary Phelps Jacob invented the bra.
- Gideon Sundback invented the modern zipper.

1914
- Garrett A. Morgan invented the Morgan gas mask.

1915
- Eugene Sullivan and William Taylor co-invented Pyrex®.

1916
- Radio tuners that received different stations were invented.
- Henry Brearly invented stainless steel.

1917
- Gideon Sundback patented the modern zipper (not the first zipper).

1918
- Edwin Howard Armstrong invented the superheterodyne radio circuit. Today, every radio or television set uses this invention.
- Charles Jung invented fortune cookies.

1919
- Charles Strite invented the pop-up toaster
- Short-wave radio was invented.
- The flip-flop circuit was invented.
- The arc welder was invented.

1920
- John T. Thompson patented the tommy gun.
- Earle Dickson invented the Band-Aid®.

1921
- Artificial life — the first robot was built.

1922
- Sir Frederick Grant Banting invented insulin.
- The first 3-D movie (requiring spectacles with one red and one green lens) was released.

1923

- Garrett A. Morgan invented a traffic signal.
- Vladimir Kosma Zworykin invented the television or iconoscope (cathode-ray tube).
- John Harwood invented the self-winding watch.
- Clarence Birdseye invented frozen food.

1924

- Chester Rice and Edward Kellogg invented the dynamic loudspeaker.
- Notebooks with spiral bindings were invented.

1925

- John Logie Baird invented the mechanical television, a precursor to the modern television.

1926

- Robert H. Goddard invented liquid-fueled rockets.

1927

- Eduard Haas III invented PEZ® candy.
- J.W.A. Morrison invented the first quartz crystal watch.
- Philo Taylor Farnsworth invented a complete electronic TV system.
- Technicolor was invented.
- Erik Rotheim patented an aerosol can.
- Warren Marrison developed the first quartz clock.
- Philip Drinker invented the iron lung.

1928

- Scottish biologist Alexander Fleming discovered penicillin.
- Walter E. Diemer invented bubble gum.
- Jacob Schick patented the electric shaver.

1929

- Paul Galvin invented the car radio.
- The yo-yo was re-invented and became an American fad.

1930

- Engineer Richard G. Drew invented Scotch® tape, patented by 3M.
- Wallace Carothers and DuPont Labs invented neoprene.

- Vannevar Bush at MIT invented the "differential analyzer" or analog computer.
- Frank Whittle and Dr. Hans von Ohain both invented a jet engine.

1931
- Harold Edgerton invented stop-action photography.
- Germans Max Knott and Ernst Ruska co-invented the electron microscope.

1932
- Edwin Herbert Land invented Polaroid® photography.
- The zoom lens and the light meter were invented.
- Carl C. Magee invented the first parking meter.
- Karl Guthe Jansky invented the radio telescope.

1933
- Edwin Howard Armstrong invented frequency modulation (FM radio).
- Stereo records were invented.
- Richard M. Hollingshead buildt a prototype drive-in movie theater in his driveway.

1934
- Percy Shaw invented cat eyes or road reflectors.
- Charles Darrow claimed he invented the game Monopoly®.
- Joseph Begun invented the first tape recorder for broadcasting the first magnetic recording.

1935
- Wallace Carothers and DuPont Labs invented nylon.
- The first canned beer was made.
- Robert Watson-Watt patented radar.

1936
- Bell Labs invented the voice-recognition machine.
- Samuel Colt patented the Colt® revolver.

1937
- Chester F. Carlson invented the photocopier.

1938
- Laszlo Biro invented the ballpoint pen.

- Strobe lighting invented.
- Albert Hofmann of Sandoz Laboratories synthesized LSD on November 16.
- Roy J. Plunkett invented tetrafluoroethylene polymer, or Teflon®.
- Nescafé® or freeze-dried coffee was invented.
- The first working turboprop engine was invented.

1939
- Igor Sikorsky invented the first successful helicopter.
- The electron microscope was invented.

1940
- Wilhelm Reich invented the Orgone Accumulator.
- Peter Goldmark invented the modern color television system.
- Karl Pabst invented the Jeep®.

1941
- Konrad Zuse's Z3 was the first computer controlled by software.

1942
- John Atanasoff and Clifford Berry built the first electronic digital computer.
- Max Adolf Mueller designed a turboprop engine.
- Robert A. Hingston invented a needle-free injector.

1943
- Synthetic rubber invented.
- Richard James invented the Slinky®.
- James Wright invented Silly Putty®.
- Albert Hofmann discovered the hallucinogenic properties of LSD.
- Emile Gagnan and Jacques Cousteau invented the aqualung.

1944
- Willem Kolff invented the kidney dialysis machine.
- Percy Lavon Julian invented synthetic cortisone.

1945
- Vannevar Bush proposed hypertext.
- The atomic bomb was invented.

1946

- Percy Spencer invented the microwave oven.

1947

- Dennis Gabor developed the theory of holography.
- Cell phones were first invented, although they were not sold commercially until 1983.
- John Bardeen, Walter Brattain, and William Shockley invented the transistor.
- Earl Silas Tupper patented the Tupperware® seal.

1948

- Walter Frederick Morrison and Warren Franscioni invented the Frisbee®.
- George de Mestral invented Velcro®.
- Robert Hope-Jones invented the Wurlitzer jukebox.

1949

- Cake mix invented.

1950

- Ralph Schneider invented the first credit card (Diners).
- John Hopps invented the first cardiac pacemaker.

1951

- Super-glue was invented.
- Francis W. Davis invented power steering.
- Charles Ginsburg invented the first video tape recorder (VTR).

1952

- Mr. Potato Head® patented.
- The first patent for bar code was issued to inventors Joseph Woodland and Bernard Silver.
- The first diet soft drink was sold.
- Edward Teller and team built the hydrogen bomb.

1953

- Radial tires were invented.
- The first musical synthesizer was invented by RCA.
- David Warren invented the "black box" flight recorder.
- Texas Instruments invented the transistor radio.

1954

- Oral contraceptives ("the pill") were invented.
- Daryl Chaplin, Calvin Fuller, and Gerald Pearson invented the solar cell.

1955

- Tetracycline was invented.
- Optic fiber was invented.

1956

- The first computer hard disk was used.
- Christopher Cockerell invented the hovercraft.
- Bette Nesmith Graham invented "Mistake Out," later renamed Liquid Paper®, to paint over mistakes made with a typewriter.

1957

- Fortran (computer language) was invented.

1958

- The computer modem was invented.
- Gordon Gould invented the laser.
- Richard Knerr and Arthur "Spud" Melin invented the Hula Hoop.
- Jack Kilby and Robert Noyce invented the integrated circuit.

1959

- Wilson Greatbatch invented the internal pacemaker.
- The Barbie® doll was invented.
- Jack Kilby and Robert Noyce both invented the microchip.

1960

- The halogen lamp was invented.

1961

- Valium was invented.
- The nondairy creamer was invented.

1962

- The audio cassette was invented.
- Yukio Horie invented the fiber-tip pen.
- *Spacewar!*®, the first computer video game, was invented.
- Dow Corporation invented silicone breast implants.

1963
- The video disk was invented.

1964
- Acrylic paint was invented.
- Permanent-press fabric was invented.
- John George Kemeny and Tom Kurtz invented BASIC (an early computer language).

1965
- AstroTurf® was invented.
- Soft contact lenses were invented.
- NutraSweet® was invented.
- James Russell invented the compact disk.
- Stephanie Louise Kwolek invented Kevlar®.

1966
- Electronic fuel injection for cars was invented.

1967
- The first handheld electronic calculator was invented.

1968
- Douglas Engelbart invented the computer mouse.
- The first computer with integrated circuits was made.
- Robert Dennard invented RAM (random access memory).

1969
- The ARPANET (first internet) was invented.
- The artificial heart was invented.
- The ATM was invented.
- The bar-code scanner was invented.

1970
- The daisy-wheel printer was invented.
- Alan Shugart invented the floppy disk.

1971
- The dot-matrix printer was invented
- The food processor was invented.
- James Fergason invented the liquid-crystal display (LCD).

- Federico Faggin, Ted Hoff, and Stanley Mazor invented the microprocessor.
- The VCR or videocassette recorder was invented.

1972
- The word processor was invented.
- John Stalberger and Mike Marshall invented the Hacky Sack.

1973
- Gene splicing was invented.
- Robert Metcalfe and Xerox invented the ethernet (local computer network).

1974
- Arthur Fry invented Post-It® notes.
- Giorgio Fischer, a gynecologist, invented liposuction.

1975
- The laser printer was invented.
- The push-through tab on a drink can was invented.

1976
- The ink-jet printer was invented.

1977
- Raymond V. Damadian invented magnetic resonance imaging.

1978
- Dan Bricklin and Bob Frankston invented the VisiCalc spreadsheet.
- Robert K. Jarvik invented the first successful permanent artificial heart, the Jarvik 7®.

1979
- Seymour Cray invented the Cray® supercomputer.
- The Walkman® was invented.
- Scott Olson invented roller blades.

1980
- The hepatitis-B vaccine was invented.

1981
- MS-DOS was invented.

- The first IBM-PC was invented.
- Gerd Karl Binnig and Heinrich Rohrer invented the scanning tunneling microscope.

1982
- Human growth hormone was genetically engineered.

1983
- The Apple® Lisa® was invented.
- The soft bifocal contact lens was invented.
- The first Cabbage Patch Kids® were sold.
- Programmer Jaron Lanier first coined the term "virtual reality."
- The 3D printer was invented.

1984
- The CD-ROM was invented.
- The Apple® Macintosh® was invented.

1985
- Microsoft® invented the Windows® program.

1986
- J. Georg Bednorz and Karl A. Muller invented a high-temperature super-conductor.
- G. Gregory Gallico, III invented synthetic skin.
- Fuji introduced the disposable camera.

1987
- The first 3D video game was invented.
- Disposable contact lenses were invented.

1988
- Digital cellular phones were invented.
- The RU-486 (abortion pill) was invented.
- Christian Andreas Doppler invented Doppler® radar.
- Ray Fuller at the Eli Lilly Company invented Prozac®.
- The first patent for a genetically engineered animal was issued to Harvard University researchers Philip Leder and Timothy Stewart.
- Ralph Alessio and Fredrik Olsen received a patent for the Indiglo® nightlight. The bluish green light is used to illuminate the entire face of a watch.

1989

- High-definition television was invented.

1990

- Tim Berners-Lee created the World Wide Web, the internet protocol (HTTP), and the WWW language (HTML).

1991

- The digital answering machine was invented.

1992

- The smart pill was invented.

1993

- Vinod Dham invented the Pentium® processor.

1994

- The HIV protease inhibitor was invented.
- The needleless syringe was patented.

1995

- The Java computer language was invented.
- The DVD (digital versatile disc or digital video disc) was invented.

1996

- Web TV was invented.

1997

- The gas-powered fuel cell was invented.

1998

- Viagra® was invented.

1999

- Scientists measured the fastest wind speed ever recorded on Earth, 509 km/h (318 mph).
- Tekno Bubbles® was patented.

Inventions of the First Decade of the 21st Century

2000

- Dr. Thottathil Varughese invented environmentally friendly transformer fluid from vegetable oil.

2001

- Kenneth Matsumaura and the Alin Foundation invented the artificial liver.
- Aprilia invented the fuel cell bike.
- PPG Industries invented self-cleaning windows.
- Jonathan Ive's invention, the Apple® iPod®, was publicly announced on October 23.
- David and Jay Groen perfected the gyroplane.
- Wilkinson Eyre Architects and Gifford and Partners, civil engineers, invented the Millennium Bridge, an alternative design to the drawbridge.

2002

- Ryan Patterson invented the Braille Glove.
- British Waterways Scotland invented the Falkirk Wheel, the world's first rotating boat lift.
- Ortho-McNeil Pharmaceutical invented the birth control patch.
- Richard Merrill invented the date-rape drug Spotter.
- Jorg Schlaich invented the solar tower.
- Canesta and VKB invented the virtual keyboard.

2003

- Alex Zettl invented the nanoscale motor, small enough to ride on the back of a virus.
- Toyota's hybrid car was introduced in Japan.
- Claudia Escobar and Skini invented salmon skin leather, used primarily in bikinis.
- Rod Sprules invented the Java Log®, made from used coffee grounds.
- Singapore Technologies Electronics and the Singapore Defense Science and Technology Agency invented the Infrared Fever Screening System, used to scan for people with high temperature or SARS in public buildings.

- Adam Whitton and Yolita Nugent invented the No-Contact Jacket®, which protects that wearer by electric shocking any attackers.

2004

- Gransdale and Alderwood invented Intel® Express Chipsets, providing built-in sound and video capabilities for the PC.
- St. George's Medical School in London invented a process to grow vaccines in plants, leading the way for cheaply manufactured vaccines.
- Robert Langer invented SonoPrep®, which injects medication by sound waves.

2005

- Steve Chen, Chad Hurley, and Jawed Karim invented YouTube®.
- Smart plastics that change shape with certain light wavelengths were invented.

2006

- The infrared alcohol test was invented to quickly identify drunk drivers.
- The Clever Car, which runs on natural gas and gets 108 mpg, was invented by BMW and various European institutions.
- Tesla Motors invented the Tesla® Roadster 100, which runs on a huge lithium-ion battery, taking it up to 250 highway miles on a single charge.
- Shanghai-based Horizon Fuel Cell Technologies invented a toy car powered by hydrogen extracted from tap water.
- Aqua Sciences invented "The Rainmaker," which harvests drinking water out of the air.
- Allerca invented a mixed-breed hypoallergenic cat.
- INSCENTINEL of Britain invented a system to train bees to detect explosives.
- HelioVolt invented Solar Skin, which can be printed directly onto glass, metal, and other building materials so skyscrapers will be solar ready.

2007

- Jonathan Ive invented the iPhone® for Apple®.
- Bruce Crower invented a steam/gasoline hybrid car that travels 40% farther than traditional gas-powered engines on a full tank.
- A British research team led by Sir Magdi Yacoub, a professor of cardiac surgery at Imperial College London and one of the world's leading heart surgeons, invented a process to grow part of a human heart from stem cells.
- The Espresso Book Machine was invented, which prints out a 300-page novel in three minutes at $3 a book.
- Sony invented the Bio Cell, a battery that runs on any sugary solution and glucose-digesting enzymes to extract electrons, creating electricity.
- Frank Pringle of Global Resource Corporation invented an emission-free process for pulling oil out of shale rock, tires, and plastic, using microwaves.
- Henry Liu invented a process to compress mercury-laced ash from coal power plants into bricks made at room temperature, also conserving energy.
- The Hyshot Scramjet Engine was invented by a team from the University of Queensland led by Professor Allan Paull. It is designed to boost a vehicle to hypersonic speeds.
- Alex Zettl invented a nanoscale radio receiver.
- The Electro Needle Biomedical Sensor Array was invented. It is a patch that takes blood readings without external extraction of blood.
- Dow Corning invented special clothing for bikers that remains soft unless impacted by a fall, at which time the entire piece of clothing becomes rigid, protecting the body.

2008

- The Italian Company Italcementi invented TX Active®, a smog-eating cement, which destroys airborne pollutants by up to 60%.
- Flying Windmills were invented by Sky Windpower, a San Diego company that extracts energy from the jet stream, possibly satisfying the planet's energy requirements.
- Students from Deakin University in Melbourne, Australia, invented the Deakin T-squared Green Car. It uses compressed air

as its fuel and is, therefore, a zero-emission vehicle. It is cheap to produce, since it uses lightweight materials, only has three wheels, and uses a "drive-by-wire" system, meaning that there is no steering rack or steering rod. A person controls the vehicle by an electronic connection rather than by a mechanical connection between the hands and steering wheel.

- Dr. Fiona Wood developed the Spray-on Skin Gun, which uses a person's own stem cells for regenerating skin tissue.
- The Nintendo® Wii®, the first interactive gaming system, was released.

2009

- The Sixth Sense, a wearable gestural interface, was invented, letting natural hand gestures interact with a computer.
- The Retinal Implant Research Group invented retinal implants, which are microelectronic implants that restore vision to patients with macular degeneration and blindness.
- Clean Seas Tuna of Port Augusta, Australia, invented the Tank-bred Tuna System.

Source: http://inventors.about.com/bio/Mary-Bellis-496.htm
Reproduced with the permission of Mary Bellis

Social Progress Index 2015

Country	Rank (SPI)	Social Progress Index	Rank BHN)	Basic Human Needs	Rank (FW)	Foundations of Well-being	Rank (O)	Opportunity
Norway	1	88.36	9	94.8	1	88.46	9	81.82
Sweden	2	88.06	8	94.83	3	86.43	5	82.93
Switzerland	3	87.97	2	95.66	2	86.5	10	81.75
Iceland	4	87.62	6	95	4	86.11	11	81.73
New Zealand	5	87.08	17	92.87	6	82.77	2	85.61
Canada	6	86.89	7	94.89	14	79.22	1	86.58
Finland	7	86.75	3	95.05	8	82.58	7	82.63
Denmark	8	86.63	1	96.03	7	82.63	12	81.23
Netherlands	9	86.5	9	94.8	5	83.81	13	80.88
Australia	10	86.42	13	93.73	12	79.98	3	85.55
United Kingdom	11	84.68	19	92.22	15	79.04	6	82.78
Ireland	12	84.66	15	93.68	29	76.34	4	83.97
Austria	13	84.45	4	95.04	9	82.53	18	75.77
Germany	14	84.04	12	94.12	10	81.5	16	76.49
Japan	15	83.15	5	95.01	20	78.78	19	75.66
United States	16	82.85	21	91.23	35	75.15	8	82.18
Belgium	17	82.83	13	93.73	27	76.57	14	78.19
Portugal	18	81.91	18	92.81	31	76.17	15	76.76
Slovenia	19	81.62	16	92.88	11	80.87	24	71.12

A full list of the study results can be found in Appendix B of the 2015 report at: **http://wikiprogress.org/articles/initiatives/ social-progress-index/#Social_Progress_Index_2015_Report**

Timeline for the Leisure Society

SOURCE: A. J. Veal, The Elusive Leisure Society (UTS, 2009)

Alfred Marshall, 1895, UK: *Principles of Economics*

The concept of a "leisure society" is not discussed, but the issue of reduced working hours is. Marshall is in favor of reduced wages, corresponding to reduced expenditure on "superfluities," but generally against these on the grounds that workers would not adapt easily to increased leisure.

Thorstein Veblen (L), 1899, USA: *The Theory of the Leisure Class*

The "leisure class" is an élite, characterized by "conspicuous leisure" and "conspicuous consumption" in middle-class households with working heads. Emulation of both practices can be carried out vicariously by wives and children but among the working class may be confined to emulation of conspicuous consumption only.

David Riesman (L), 1950 ,USA: *The Lonely Crowd*

Riesdman initially suggests that leisure might offer scope for autonomy and meaning as these disappear from industrial working life, but he later rejects this view, calling for the humanization of both work and leisure.

Jacques Ellul, 1954, France: *The Technological Society*

Leisure is no solution to alienating work, since leisure is also controlled by the industrial system, and any attempt to exercise real freedom in leisure would be frustrated by the system.

George Soule, 1955, USA: *Time for Living*

"Democratic leisure" is seen as potentially destructive but is also put forward as part of a "new solution" to reduced working hours.

Herbert Marcuse, 1955, USA: *Eros and Civilization*

Marcuse echoes Ellul in arguing that leisure is controlled by the industrial system rather than being a realm of freedom.

John Kenneth Galbraith, 1967, USA: *The New Industrial State*

Under the current conditions of "demand management" by the industrial system, the evidence suggests that increasing productivity (and wage rates) will not result in more leisure but instead in more consumption.

John Tribe, 2005, UK: *Trends in Work and Leisure: A Leisure Society?*

The idea of a leisure society involves a number of paradoxes, and while the economic circumstances exist for a leisure society, individual and political factors have resulted in society being more rather than less work-orientated.

Bibliography & Online Resources

Aaron, J. Henry, and William B. Schwartz (Eds.). *Coping with Methuselah – The Impact of Molecular Biology on Medicine.* Washington D.C.: Brooking Institution Press, 2003.

Adams, James Truslow. *The Epic of America.* New York: Simon Publications, 2001.

Aristotle. *Ethics.* Edited and translated by John Warrington. Dutton, NY: Everyman's Library, 1975.

Asimov, Isaac. *Chronology of Science and Discovery.* New York: Harper and Row Publishers, 1989.

Asimov, Isaac. *The Rest of the Robots.* New York: Doubleday Publishing Company, 1964.

Babbage, Charles. *The Exposition of 1851.* London, UK: Cass, 1968.

Bacon, Francis. *The New Organon* (Cambridge Texts in the History of Philosophy). Cambridge, UK: Cambridge University Press, 2000.

Barrat, James. *Artificial Intelligence and the End of the Human Era.* New York: St Martin's Press, 2015.

Bostrom, Nick. *Superintelligence – Paths, Dangers, Strategies.* Oxford, UK: Oxford University Press, 2014.

Bowen, Huw. *The Business of Empire: The East India Company and Imperial Britain, 1750–1833.* Cambridge, UK: Cambridge University Press, 2006.

Burke, James, and Robert Ornstein. *The Axemaker's Gift.* New York: Tarcher Perigee, 1997.

Carson, Rachel. *The Silent Spring* (Anniversary edition). Evanston, IL: Houghton Mifflin Company, 2002.

Christian, David. *Maps of Time: An Introduction to Big History.* Berkeley, CA: University of California Press, 2004.

Cooper, Robert. *The Breaking of Nations: Order and Chaos in the Twenty-First Century.* London, UK: Atlantic Books, 2004.

Coyne, A. Jerry. *Why Evolution Is True.* New York: Brilliance Publishing, 2012.

de Grey, Aubrey, and Michael Rae. *Ending Aging: The Rejuvenation Breakthroughs That Could Reverse Human Aging in Our Lifetime.* New York: St Martin's Griffin, 2008.

de Soto, Hernando. *The Mystery of Capital: Why Capitalism Triumphs in the West and Fails Everywhere Else.* New York: Basic Books, 2000.

Diamond, Jared. *Guns, Germs and Steel:* New York: W.W. Norton and Company, 1997.

Dosi, Giovanni. *Innovation, Organization and Economic Dynamics: Selected Essays.* London: Edward Elgar Publishing, 2000.

Dressler, K. Eric. *Engines of Creation: The Coming Era of Nanotechnology.* New York: Anchor, 1987.

Findlay, Ronald, and Kevin O'Rourke. *Power and Plenty: Trade, War and the World Economy in the Second Millennium.* Princeton, NJ: Princeton University Press, 2007.

Foot, Philippa. *Virtues and Vices.* Oxford, UK: Clarendon Press. 2012.

Freeman, Michael. *Human Rights.* Malden, MA: Polity Press, 2002.

Grayling, A. C. *Ideas That Matter: A Personal Guide for the Twenty-First Century.* London: Orion, 2009.

Green, Brian. *The Fabric of the Cosmos: Space, Time and The Texture of Reality.* New York: Vintage Books, 2004.

Harper, Sarah, and Hamblin Kates (Eds.). *International Handbook on Ageing and Public Policy.* London, UK: Edward Elgar Publishing, 2014.

Kelly, Kevin. *What Technology Wants.* New York: Penguin Books, 2010.

Kitcher, Philip. *The Ethical Project*. Cambridge, MA: Harvard University Press, 2014.

Kurzweil, Ray. *The Age of Spiritual Machines*. New York: Penguin Books, 1999.

Kurzweil, Ray. *The Singularity Is Near*. New York: Viking Penguin, 2005.

McLuhan, Marshall. *The Gutenberg Galaxy*. Toronto: University of Toronto Press, 1962.

Mills, C. Wright. *The Sociological Imagination*. New York: Oxford University Press, 1959.

Morris, Jan. *Heaven's Command: An Imperial Progress*. New York: Mariner Books, 2002.

Mumford, Lewis. *The Lewis Mumford Reader*. New York: Pantheon Books, 1986.

Le Guin, Ursula K. *The Wave in the Mind: Talks and Essays on the Writer, the Reader and the Imagination*. Boston, MA: Shambhala Publications, 2004.

Naam, Ramez. *The Infinite Resource: The Power of Ideas on a Finite Planet*. Lebanon, NH: University Press of New England, 2013.

Newton, Isaac. *The Three Laws of Motion*. Knoxville, TN: University of Tennessee.

Parthasarathi, Prasannan. *Why Europe Grew Rich and Asia Did Not: Global Economic Divergence, 1600–1850*. Cambridge, UK: Cambridge University Press, 2011.

Pinker, Steven. *The Better Angels of Our Nature: Why Violence Has Declined*. New York: Penguin Books, 2012.

Pommeranz, Kenneth. *The Great Divergence: China, Europe and the Making of the Modern World*. Princeton, NJ: Princeton University Press, 2001.

Radjou, Navi, Jaideep Pradbhu, and Simone Ahuha. *Think Frugal, Be Flexible, Generate Breakthrough Growth*. San Francisco, CA: Jossey-Bass, 2012.

Ravenhill, John. *Global Political Economy*. New York: Oxford University Press, 2005.

Reese, Byron. *Infinite Progress – How the Internet and Technology Will End Ignorance, Disease, Poverty, Hunger and War*. Austin, TX: Greenleaf Book Group, 2013.

Reese, Martin. *Our Final Century: The 50/50 Threat to Humanity's Survival*. Toronto: Penguin Random House Canada, 2004.

Riello, Georgio, and Prasannan Parthasarathi. *The Spinning World: A Global History of Cotton Textiles, 1200–1850*. Oxford, UK: Oxford University Press, 2009.

Rifkin, Jeremy. *The Third Industrial Revolution*. New York: St Martin's Griffin, 2013.

Rosenau, M. James. *Distant Proximities: Dynamics Beyond Globalization*. Princeton: Princeton University Press, 2003.

Sagan, Carl. *The Demon Haunted World: Science as a Candle in the Dark*. New York: Ballantine Books, 1997

Sagan, Carl. *The Dragon of Eden: Speculations on the Evolution of Human Intelligence*. New York: Ballantine Books, 1978.

Schmidt, Eric, and Jared Cohen. *The New Digital Age: Reshaping the Future of People, Nations and Business*. New York: John Murray, 2013.

Schwartz, Barry. *The Paradox of Choice. Why More Is Less. How the Culture of Abundance Robs Us of Satisfaction*. New York: Harper Collins Publishers, 2005.

Sen, Amartya. *The Idea of Justice*. Cambridge, MA: Belknap Press, 2011.

Sen, Amartya Kumar. *Poverty and Famines: An Essay on Entitlement and Deprivations*. Oxford, UK: Oxford University Press, 1981.

Smil, Vaclav. *Transforming the Twentieth-Century: Technical Innovations and Their Consequences*. New York: Oxford University Press, 2006.

Smith, Adam. *An Inquiry into the Nature and Causes of the Wealth of Nations: A Selected Edition*. Oxford, UK: Oxford University Press, 1993.

Stiglitz, Joseph E. *The Great Divide: Unequal Societies and What We Can Do About Them*. New York: W.W. Norton & Company, 2015.

Taleb, N. Nassim. *The Black Swan: The Impact of Highly Improbable Fragility* (2nd ed.). New York: Random House Publishing Group, 2010.

Teilhard de Chardin, Pierre. *The Phenomenon of Man*. New York: Harper Perennial Modern Classics, 2008.

Turkle Sherry. *Alone Together: Why We Expect More from Technology and Less from Each Other*. New York: Basic Books, 2012.

Turkle, Sherry. *Reclaiming Conversation: The Power of Talk in a Digital Age*. New York: Penguin Books, 2015.

Wallerstein, Immanuel. *The Modern World System. Book 2: Mercantilism and the Consolidation of European World Economy, 1600–1750*. Berkeley, CA: University of California Press, 2011.

Weber, Max. *The Protestant Ethic and the Spirit of Capitalism*. London, UK: Talcott Parsons, 1982.

Wright, Robert. *NONZERO: The Logic of Human Destiny*. New York: Pantheon Books, 1999.

Online Resources

Abilov, Shamkhal, 2011. "Industrial Revolution and the Great Divergence between the East and the West." http://www.academia.edu/13142228/The_Great_Divergence_Between_China_and_England_Why_Industrial_Revolution_Happened_in_Europe

Allen G.P., Olveda M. Remigio, Yuesheng Li. 2014. "Are We Ready for a Global Pandemic of Ebola Virus?" www.sciencedirect.com/science/article/pii/S1201971214016178

Barzum, Jacques. "Revolution and the Growth of Industrial Society, 1789–1914." *Encyclopedia Britannica*, Sections 33 to 43. http://www.britannica.com/topic/history-of-Europe/Revolution-and-the-growth-of-industrial-society-1789-1914

Bostrom, Nick. 2007. "The Future of Humanity" www.nickbostrom.com/papers/future.pdf

Brain. 2025. "A Scientific Vision" http://braininitiative.nih.gov/pdf/BRAIN2025_508C.pdf

Brink, Lindsay. 2011. "Frontier Economics" www.the-american-interest.com/2011/09/28/frontier-economics

Burke, James, and Robert Ornstein. *The Human Journey: A Report on The Axemaker's Gift*. www.humanjourney.us/axemaker.html

De Waal, Frans. "The Bonobo in All of US." NOVA www.pbs.org/wgbh/nova/nature/bonobo-all-us.htm

Fresco, Jacques. "The Venus Project: Beyond Poverty Politics and War" https://www.thevenusproject.com/the-venus-project/

Hathaway, Ian, and Robert E. Litan. "Declining Business Dynamism in the United States: A Look at States and Metros" www.brookings.edu/research/papers/2014/05/declining-business-dynamism-litan

Howse, Robert. "Montesquieu on Commerce, Conquest, War, and Peace" http://www.law.nyu.edu/sites/default/files/ECM_PRO_060042.pdf

International Diabetes Federation. 2009. "Latest Diabetes Figures Paint Grim Global Picture" www.idf.org/latest-diabetes-figures-paint-grim-global-picture

Kevin, Kelly. "Singularity is Always Near" http://kk.org/thetechnium/the-singularity/

Kirkwood, B.L. Thomas. 2015. "Are We Designed to Die" www.biochemist.org/bio/03704/0008/037040008.pdf

Magee, C.L., and T.C. Devezas. "How Many Singularities Are Near and How Will They Disrupt Human History?" http://web.mit.edu/~cmagee/www/documents/29-singularitysdarticle.pdf

Muneeruddin, Hinasahar. "The Three Agricultural Revolutions" http://www.lewishistoricalsociety.com/wiki/tiki-print_article.php?articleId=2

Nano Medicine. "Health Care in the 21th Century" www.thenanoage.com/nanomedicine.htm.

National Human Genome Institute. "Cloning" https://www.genome.gov/25020028/cloning-fact-sheet/

National Institute of General Medical Sciences. "The New Genetics" https://publications.nigms.nih.gov/thenewgenetics/chapter1.html

New World Encyclopedia. "Mercantilism" www.newworldencyclopedia.org/entry/Mercantilism

OECD. "Promises and Perils of a Dynamic Future. OECD Report 35391210: 21st Century Technology" http://www.oecd.org/futures/35391210.pdf

Palme, Jacob. 2012. "The Future of Homo Sapiens – The Future of Human Evolution" futurehumanevolution.com/the-future-of-homo-sapiens

Pew Research. 2014. "US Views of Technology and the Future" www.pewinternet.org/2014/04/17/us-views-of-technology-and-the-future/

Shelton, A.M. "Are Organic Agriculture and Biotechnology Compatible?" www.agribiotech.info/.../Biotech%20and%20Organic%20final%2004%20layout.pdf

Stanford Encyclopedia of Philosophy. 2008. "Ethics of Stem Cell Research." plato.stanford.edu/entries/stem-cells/

Weitzner, Daniel. 2015-07-06. "Keys Under Doormats: Mandating Insecurity by Requiring Government Access to All Data and Communications"
http://hdl.handle.net/1721.1/97690

Wikipedia. "Transhumanism"
https://en.wikipedia.org/wiki/Transhumanism

Williams, R.B., B.M. Jenkins, and D. Nguyen. "Solid Waste Conversion: A review and database of current and emerging technologies"
http://biomass.ucdavis.edu/files/2013/10/10-16-2013-2003-solid-waste-conversion-review-and-assessment.pdf

World Economic Forum. "Five Better Indicators than GDP."
https://www.weforum.org/agenda/2016/04/five-measures-of-growth-that-are-better-than-gdp/

Index

A

Abilov, Shamkhal, 20

Adams, James Truslow, 92
    American Dream, 92, 122, 126

Age of Exploration, 19, 125

Age of Reason, 89–90

aging, 69–73, 211–12, *See also* life
    expectancy

agriculture, 30, 36, 43, 56, 63–68,
    67f, 88, 92, 99, 130–31, 131f, 203,
    219

Ahuja, Simone, 62, 126
    Jugaad Innovation, 62

Albert, Prince, 121

Alphabet conglomerate, 72

Amazon, 127

American Dream, 92, 122, 126

Amero (currency), 167

Anderson, Sanders, 115

Anti-Trust Act (US), 94

Arab Spring, 106, 179

Archimedes, 190

Aristotle, 22, 38, 106, 175

ARPANET, 33, 209

artificial intelligence (AI), 13, 37,
    56, 59, 114–16, 156, 192, 195,
    199–201, 204, 206
    artificial general intelligence
        (AGI), 200–01
    artificial superintelligence (ASI),
        114–16, 200–01
    narrow artificial intelligence
        (NAI), 200

Asimov, Isaac, 16–17, 18, 33, 55,
    114, 158, 220
    *The Chronology of Science and
        Discovery,* 16, 17, 55
    laws of robotics, 114, 220

Ayanz y Beaumont, Jeronimo de,
    87

B

Babbage, Charles, 28, 29, 205–06

Bacon, Francis, 24–25, 89–90, 194
    Novum Organum, 25

Banting, F.G., 33, 208

Bardeen, John, 33

Barrat, James
    *Artificial Intelligence and the End
        of the Human Era,* 115

Bell, Alexander G., 28, 30

Bellis, Mary, 21

Benz, K., 33

Berners-Lee, Tim, 33, 210

Bessemer, Henry, 28
    Bessemer converter, 28, 95

Best, C.H., 33, 208

Better Life Index (BLI), 152

Bible, 23, 25

Big Bang theory, 109, 161, 204, 216

biotechnology, 35, 56, 65, 99, 114,
    156, 204, 211

black swan concept, 109–10, 114

Boon, Mary E., 39

Bostrom, Nick, 201

BRAIN, 114, 201, 204

Brain Activity Mapping (BAM), 115–16, 155

Branca, Giovana, 22

Brattain, Walter H., 33, 220–21

Brito, Dagobert L., 129

Burke, James, 89, 190
  The Axemaker's Gift, 89

C

Calico (company), 72

Cantillon, Richard, 135

capitalism, 60, 121, 176–78

carbon dioxide emissions. See greenhouse gases

cars, 29, 31, 95, 130
  driverless cars, 10, 36, 99, 200
  electric cars, 36, 38, 74, 202

Carson, Rachel
  The Silent Spring, 157, 183–84

Cartwright, Edmund, 26

Century of Reason, 26

Cerf, Vincent, 209

Chaplin, Charlie
  Modern Times, 96

Chekhov, Anton, 165

child labor, 88, 93–94, 148

Christian, David, 44
  The History of our World in Eighteen Minutes, 44

Churchill, Winston (UK PM), 149

Civil War (US), 94, 122

Clark, Arthur C., 156

Clausius, Rudolph, 202

clean air acts, 96, 184

climate change. See environment, greenhouse gases, global warming

Clinton, Bill (US pres.), 140

clock. See mechanical clock

cloning, 161, 192, 200, 202–03, 212
  recombinant DNA cloning, 203
  reproductive cloning, 203
  therapeutic cloning, 203

Colantonio, Sophia, 157

commercialization, 28, 30, 99, 122–30

complexity theory, 38–39

computers, 8, 9, 28, 29, 31, 33, 34, 39, 40f, 56, 95, 129, 151, 205–06, 207, 210

Comte, Auguste, 194

connectivity. See internet

consumer price index (CPI), 131

Cooper, Martin, 33

Copernicus, Nicolai, 21, 22–23, 190
  On the Revolution of Heavenly Bodies, 22–23

corn production, 99, 131

cosmic time scale, 50, 204

cotton, 51, 65, 66, 67f, 84, 120
  Bt cotton, 66–68, 67f
  cotton gin, 47
  cotton mills, 87

Cruickshank, George
  British Bee Hive, 90

Cugnot, Nicolas-Joseph, 27

Curl, Robert F., 129

D

Davy, H., 28

DDT, 157, 184

de Colmar, Thomas, 205

de Grey, Aubrey, 70, 211
  Strategies for Engineered Negligible Senescence (SENS), 70, 211

de Magalhães, João Pedro, 71–72, 211–12

Ethical Perspectives in Biogerontology, 71–72

de Soto, Hernando, 176–78
    The Mystery of Capital, 176–77

Denis, Jean-Baptiste, 22, 23–24

depression(s), economic, 45–46, 47
    Great Depression, 31, 45–46
    Long Depression (LD), 45–46

Descartes, René, 21, 24, 25

Diamond, Jared
    Guns, Germs, and Steel, 19

Diderot, Denis, 27, 39, 205
    Encyclopedia, 27, 39, 205

Digital Age, 38, 78, 179

Digital Revolution. *See* Third Industrial Revolution

disintermediation, 127–28

Dittmar, J., 53–55, 54f

Dosi, Giovanni
    Innovation, Organization and Economic Dynamics, 137

Drebbel, Cornelius, 22

Drexler, K. Eric
    Engines of Creation, 50, 57

E

education/training, 37, 78–80, 93, 97, 104–05, 128–29, 138, 141–42, 153t, 160, 165, 174–75, 178, 196, 204, 222
    massive open online course (MOOC), 78

Einstein, Albert, 25, 32, 190

electricity, 37, 39, 40f, 74, 75–77, 94–95, 130, 202, 220
    electric car, 36, 38, 74, 202
    electric light, 28, 29, 40f, 93
    G. Ohm, 28
    H. Davy, 28

emerging technologies, 11, 56–60, 98, 114–16, 165, 185–86, 195, 204

encyclopedia, 26, 27, 40f, 205

energy, 29, 38, 73–78, 73f, 74f, 96, 98, 134–35, 185–86, 202
    nuclear power, 73f, 74f, 75
    renewable energy, 38, 73f, 74–77, 185–86
    waste to energy, 76, 77

Enlightenment period, 26, 27, 55, 68, 205

entrepreneurship, 92, 104, 132, 135–38

environment, 63, 75–77, 96, 151, 157, 166, 169, 182–86

Epic of Gilgamesh, 210

Escagenetics (company), 99

ethics/ethical issues, 66, 71–72, 148, 158–61, 192, 203, 211–12

Ewing Marion Kauffman Foundation, 136–37, 136f

F

Facebook, 36, 40f, 105–06, 110, 172t

factory system, 30, 51–53, 55, 84–85, 88–90, 92–93, 96, 129, 213

Faraday, Michael, 190

farming. *See* agriculture

Farnsworth, P., 33

Fleming, A., 33

Florida, Richard, 139–40
    Global Creativity Index (GCI), 139–40, 139f

Fogel, Robert, 112

Ford, Henry, 33, 99, 130

Francis (pope), 161

Franklin, Benjamin, 27

free-market system, 20, 47, 62, 140, 151, 168–69, 177

Freitas, Robert, Jr., 69–70

French Revolution, 90

Fresco, Jacque
  Venus Project Society, 167

Freud, Sigmund, 181

Froelich, John, 28

frugal innovation, 62–63, 126

fuel cells, 28, 29, 202
  William R. Grove, 28, 202

Fyodorov, Nikolai, 194

G

Galileo, 22–23, 27, 110, 190

gas lighting, 26, 88
  William Murdoch, 27

GDP (gross domestic product), 46, 48, 140–42, 142f, 149–52, 150t, 154t

General Electric (GE) Corp., 125, 128–29
  Vscan, 125–26

genetically modified food (GMF), 66

Giordano, Bruno, 148

Gips, James, 115

Global Challenges Foundation, 115

Global Creativity Index (GCI), 139–140, 139f

global warming, 75–76, 96, 183

globalization, 20, 62, 78, 91, 99, 111, 124–27, 149, 166, 169

Goddard, R.H., 33

Godwin, William, 194

Goodyear, C., 28, 29

Google, 36, 72, 99, 200

Gore, Albert (US vice-pres.)

The Future: Six Drivers of Global Change, 138, 184

Gould, Stephen J., 221

Great Divergence, 18–20
  Samuel P. Huntington, 18

Great Exhibition(s), 29, 90, 119–20

Greene, Bryan, 32
  The Fabric of the Cosmos, 32

greenhouse gases, 65, 74–78, 183–86, 185f, 202

Greenpeace, 182

grid parity, 76–77

GRIN, 204

Gross National Happiness (GNH), 152
  GNH Report, 152

Grove, William R., 28, 202

Guillaume, C.E., 28

gun powder, 16, 17–18, 83–84, 156
  Song dynasty (China), 84

Gutenberg, Johannes, 8, 16, 32, 86, 110, 213–14
  printing press, 8, 16, 23, 27, 32, 39, 40f, 53–55, 53f, 54f, 86–87, 110, 213–14

H

Hammarskjold, Dag, 196

Happy Planet Index (HPI), 152

Hargreaves, James, 26

Harvey, William, 24

Hawking, Stephen, 115

Hayek, Friedrich, 90

health, 56–58, 65–66, 68–73, 77, 95–96, 102f, 129, 151, 153t, 155, 166, 171, 173

heart, artificial, 33, 58, 199
  Robert Jarvik, 33

total artificial heart (TAH), 199

Heylighen, Francis, 109

Hingston, Robert A., 33

Hobbes, Thomas, 63–64

Hubbard, Elbert, 104

Homo erectus, 192

Homo habilis, 192

Homo sapiens, 57, 108, 190, 203

Hugo Victor, 130

Hull, Chuck, 33

Human Brain Project, 201

human rights, 169, 178–82

Huntington, Samuel P., 18

I

I, Robot (movie), 116

IBM, 34, 130

immigration, 90, 92, 119–20, 122, 138

industrial revolution(s), 12, 19, 22, 26, 37–38, 55, 73, 88–90, 121–23, 128–29, 156, 190–92, 219
    First Industrial Revolution, 26, 37, 38, 74, 84–85, 88, 89, 96, 184, 191, 213, 219
    Second Industrial Revolution, 12, 37, 38, 90, 122–23, 191, 219
    Third Industrial Revolution, 37, 38, 95, 192
    Fourth Industrial Revolution, 37, 38, 128–29, 192

informatics, 13, 114, 156, 204

information technology, 20, 34–38, 47, 56, 58, 78, 105–06, 127–28, 134–35, 142–43, 170, 172, 179, 204–05

insulin, 33, 95, 203, 208
    C.H. Best, 33, 208
    F.G. Banting, 33, 208

International Day of Happiness, 152

International Exhibition (Paris), 119

International Technology Education Association (ITEA), 13

International Telecommunication Union (ITU), 172
    The State of Broadband, 172

internet, 32–34, 40f, 56, 61–62, 95–98, 105–11, 107f, 127–29, 170–72, 171f, 172f, 204–05, 209–10
    Big Data Exchange, 128
    Industrial internet (Ii), 128–29
    Internet of Things (IoT), 128

Invar, 28, 29–30
    C.E. Guillaume, 28

iPhone, 36, 40f

iPod, 36, 37, 40f

iTunes, 36, 40f, 111

J

Jarvik, Robert, 33

Jefferson, Thomas (US pres.), 161

Jenal, Marcus, 38–39

Jenner, Edward, 27, 31, 88

Joy, Bill, 115

Juvenal (poet), 109

K

Kahn, Bob, 209

Kardashev, Nicolai, 76

Kelly, Kevin, 12, 38
    Seventh Kingdom, 12
    What Technology Wants, 12

Kepler, Johannes, 89

Keynes, John Maynard, 142–43, 197

Economic Possibilities for our Grandchildren, 143
The General Theory of Employment, Interest and Money, 142
King, Karen
  Global Creativity Index (GCI), 139–40, 139f
Kovarik, Bill, 183
Kurzweil, Ray, 49, 59, 108, 115, 156, 194
  The Singularity is Near, 49, 59, 115, 156, 194
Kyoto Accord, 182–83

L
Labor Movement, 88, 94
Landsteiner, Karl, 24
Laurier, Wilfrid (Canada PM), 140
Law of Accelerating Returns (LOAR), 49
Le Guin, Ursula K.
  The Wave in the Mind, 12, 105
le Rond d'Alembert, Jean, 205
  Encyclopedia, 27, 39, 205
Leibniz, G.W., 22
leisure, 73, 143, 197
  leisure preference, 52–53
  leisure society, 73, 143, 197
Leopold, Aldo
  A Sand County Almanac, 184
Lesage, Georges Louis, 27
Libertarian paternalism, 180-81
Licklider, Joseph C.R., 209
life expectancy, 68–71, 69f, 87–88, 95, 111–114, 112f, 113f, 152, 155, 210–12, 215
Lincoln, Abraham (US pres.), 107
Lipperhey, Hans, 22

literacy/illiteracy, 55, 86, 97, 106, 176
longevity. *See* life expectancy

M
Maass, Alan, 60
Mach, Ernst, 32
Marconi, G., 33
Marglin, S., 52–53
  leisure preference, 52–53
Marshall Plan, 31
Martin, James, 49

Marx, Karl, 60, 124
  Das Kapital, 124
McCulloch, Ernest, 33
McFarlane, Alan
  The Riddle of the Modern World, 19, 126
McLuhan, Marshall
  The Gutenberg Galaxy, 51, 86
mechanical clock, 17–18, 52–53, 83, 85–86, 213
Medawar, Peter, 211
Mellander, Charlotta
  Global Creativity Index (GCI), 139–40, 139f
meritocracy principle, 92–93, 122
Meucci, Antonio, 30
Middle Ages, 14, 16, 64–65
Minsky, Marvin, 57
Montesquieu (political philosopher), 19, 27, 126
Moore's Law, 49, 96, 221
morphological freedom (MF), 182
Morse, Samuel, 28, 29, 90
Mozilla Firefox, 36
Mumford, Lewis

The Lewis Mumford Reader, 85
Murdoch, William, 27
Mushet, Robert F., 28

N
nanomedicine. *see* nanotechnology
nanotechnology, 13, 35–36, 50,
    56–57, 100–03, 100f, 101t, 102f,
    103f, 114–16, 156, 185, 195, 204
    nanobots, 101
Newcomen, Thomas, 87, 218
Newton, Isaac, 21, 25, 32, 89–90,
    110, 190
    Universal Laws, 21, 25, 214
NIBC, 204
nylon, 46

O
Obama, Barack (US pres.), 115
Ohm, G., 28
Olshansky, S. Jay, 72
Omega Point, 194, 195
Ornstein, Robert E., 89, 190
    The Axemaker's Gift, 89

P
pandemic disease, 169, 172–74, 208
Paré, Ambroise, 24
Paris Accord, 75, 183, 185
Pascal, Blaise, 21
Pasteur, Louis, 50
Peeters, Peter, 217–18
Pemberton, J., 28
Pension Act (US), 94
Pillans, James, 78
Pinker, Steven
    The Better Angels of Our Nature,
    148
Polaroid, 46

polio, 31, 33, 155, 208
    Jonas Salk, 33, 208
polycentrism, 126
Popper, Karl, 32
population, 16, 65, 68–69, 69f, 111,
    112f, 113f, 121–25, 153, 157,
    171t, 173, 181, 189, 207, 207t
    growth, 88, 91–92, 110, 124–25,
    131–32, 167, 215–16
    overpopulation, 71
    urban, 92f
potato production, 130–31, 131t
poverty, 148–49, 166, 174–78
    have-nots, 60–63, 165
    reduction of, 169, 174–78
Prabhu, Jaideep, 62, 126
    Jugaad Innovation, 62
Priestly, Joseph, 15
privacy, 108, 149, 178–81, 192
productivity, 49, 51, 104–05,
    130–44, 133f, 134f, 197, 208, 213,
    215
    corn, 99, 131
    labor, 132, 132f
    potatoes, 130–31, 131t
    total factor productivity (TFP),
    131, 135
Pulsifer, Catherine, 161

Q
Quelch, John, 97

R
Rabelais, François, 86
Radjou, Navi Radju, 62, 126
    Jugaad Innovation, 62
recession(s), economic, 45, 46–47,
    48, 218
    Great Recession (2007), 45, 134,
    137, 141, 149, 151
Rees, Martin
    Our Final Century, 101

religion/religious beliefs, 9, 16, 20, 23–25, 27, 59, 86, 110, 158, 174, 183, 193–95, 205, 211

Renaissance, 8, 14, 16–20, 24–25, 83–84, 194

resource-based economy (RBE), 167

Rifkin, Jeremy, 99

robots/robotics, 37, 50, 57, 59, 99, 101, 114–15, 143, 156, 192, 204, 219–20
  laws of robotics, 114, 220
  nanobots, 101

Roman Empire, 155

Roman Inquisition, 23

Rousseau, Jean-Jacques, 27, 89–90

Royal Society of London, 26

Russell, Bertrand, 148, 154–55, 158

S

Sagan, Carl, 26, 191, 204

Salk, Jonas, 33, 208

Sandberg, Anders, 182

Savery, Thomas, 22, 87, 218–19

Schmidt, Eric
  The New Digital Age, 111, 210

Schwab, Klaus, 37

Scientific Revolution, 19, 21, 55

Sen, Amartya
  The Idea of Justice, 168
  Poverty and Famines, 175

Sethe, Sebastian
  Ethical Perspectives in Gerentology, 71–72

Shakespeare, William
  Hamlet, 196

Shih Huang Ti (emperor), 18

Shockley, William B., 33, 190, 220

Siemens AG, 129

Silk Road, 124–25

Singapore, 139t, 141–42, 141f, 142f, 174

Singer, I., 28, 29

singularity concept, 49, 59, 98, 108–10, 195, 214–18, 215f

smallpox, 27, 31, 87–88, 155
  vaccination, 27, 31, 87–88

smartphones, 36, 40f, 49, 96, 98, 106, 171

Smil, Vaclav, 29
  period of synergy, 29

Smith, Adam, 60, 121–22, 168–69, 177
  The Wealth of Nations, 168–69

Snowden, David J., 38–39

social media, 8, 34, 37, 105–06, 111
  See also Facebook

Social Progress Index, 140, 152–54, 153t, 154t, 239t

Song dynasty (China), 84, 213

space probes
  Pioneer 10 and 11, 35
  Voyageur 1 and 2, 35

spinning wheel, 17–18, 26, 51f, 52, 83–85, 89

steam engine, 22, 26–27, 30, 37, 40f, 47, 55–56, 56f, 87, 90, 124, 130, 218–19
  Giovana Branca, 22
  James Watt, 26, 55, 87, 219
  Miner's Friend, 87
  Thomas Newcomen, 87, 218
  Thomas Savery, 22, 87, 218–19

Steinhart, Eric, 194

stem cells, 33, 159–60, 192, 203, 219
  Ernest McCulloch, 33
  J.E. Till, 33

Stent, Charles T., 28, 48

Stiglitz, Joseph
  The Great Divide, 94
Stimson, Henry L., 181
Strickland, Jonathan, 209
structural unemployment, 99, 104, 127, 129
Sturgeon, W., 28
Sunstein, Cass R., 180

T
Taleb, Nassim Nicholas, 110
  black swan robust society, 110
Taylor, Frederic, 31, 53, 85, 92, 122
Teilhard de Chardin, Pierre, 194
telegraph, 27, 28, 29, 40f, 90–91, 94, 123
  Georges Louis Lesage, 27
  Samuel Morse, 28, 29, 90
telescope, 22, 24, 40f
  Hans Lipperhey, 22
Thaler, Richard H., 180
Thomas Aquinas, Saint, 175–76
Thoreau, Henry David
  Walden, 183–84
Till, J.E., 33
Tithonus syndrome, 70
total factor productivity (TPF), 131, 135
transhumanism, 25, 182, 194–96
transistors, 15, 33–34, 39, 40f, 49, 95–96, 206, 220–21
  John Bardeen, 33
  Walter H. Brattain, 33, 220
  William B. Shockley, 33, 190, 220
Truman, Harry S. (US pres.), 158
Turing, Alan M., 206
  Turning test, 206
Turkle, Sherry
  Alone Together, 106
  Reclaiming Conversation, 106

Tverberg, Gail, 76
Twain, Mark
  Gilded Age, 93

U
Ulam, Stansilaw, 59
United Nations Framework Convention on Climate Change (UNFCCC), 75, 183
Universal Declaration of Human Rights, 178–79, 182
Urban, Tim, 201
USB flash drive, 36

V
vaccines, 27, 31, 33, 68, 87–88, 95, 173, 208
  Edward Jenner, 27, 31, 88
  Jonas Salk, 33, 208
Vallée, Jacques, 217
Van Dulken, Steve, 37
Venus Project Society, 167
Vesalius, Andrea, 24
Vinge, Vernor, 216–17
Volta, A., 28
Voltaire (philosopher), 27, 89–90
von Newman, John, 59, 108

W
Wagner, Richard, 63
Wallis, Stewart, 151
Watt, James, 26, 55, 87, 219
Weisman, Alan
  The World Without Us, 65
Wikipedia, 62, 135–36, 202, 205, 209
Wilhelm, Carl, 15
Wilson, John, 121

Wolfram, Stephen, 38
   complexity theory, 38–39
Woodcock, George
   The Tyranny of the Clock, 53, 86,
     213
World Economic Forum, 37, 151
   Klaus Schwab, 37
   Stewart Wallis, 151
world wars, 135
   World War I, 31, 121, 218
   World War II, 31, 121, 134, 157,
     184

World Wide Web, 33, 40f, 209–10
   Tim Berners-Lee, 33, 210
Wren, Christopher, 22

Y

Y2K, 9, 35
YouTube, 36, 40f
Yung-lo (emperor), 18

Z

Zuse, K., 33